Universitext

Springer Science+Business Media, LLC

BOOKS OF RELATED INTEREST BY SERGE LANG

Introduction to Linear Algebra, Second Edition
1986, ISBN 0-387-96205-0

Linear Algebra, Third Edition
1987, ISBN 0-387-96412-6

Undergraduate Algebra, Second Edition
1990, ISBN 0-387-97279-X

Undergraduate Analysis, Second Edition
1997, ISBN 0-387-94841-4

Complex Analysis, Third Edition
1993, ISBN 0-387-97886-0

Math Talks for Undergraduates
1999, ISBN 0-387-98749-5

OTHER BOOKS BY LANG PUBLISHED BY
SPRINGER-VERLAG

Math! Encounters with High School Students • The Beauty of Doing Mathematics •
Geometry: A High School Course • Basic Mathematics • Short Calculus • A First
Course in Calculus • Calculus of Several Variables • Algebra • Real and Functional
Analysis • Introduction to Differentiable Manifolds • Fundamentals of Differential
Geometry • Algebraic Number Theory • Cyclotomic Fields I and II • Introduction to
Diophantine Approximations • $SL_2(\mathbf{R})$ • Spherical Inversion on SLn($\mathbf{R}$) (with Jay
Jorgenson) • Elliptic Functions • Elliptic Curves: Diophantine Analysis •
Introduction to Arakelov Theory • Riemann-Roch Algebra (with William Fulton) •
Abelian Varieties • Introduction to Algebraic and Abelian Functions • Complex
Multiplication • Introduction to Modular Forms • Modular Units (with Daniel
Kubert) • Introduction to Complex Hyperbolic Spaces • Number Theory III • Survey
on Diophantine Geometry • Collected Papers I–V, including the following:
Introduction to Transcendental Numbers in volume I, Frobenius Distributions in
GL2-Extensions (with Hale Trotter) in volume II, Topics in Cohomology of Groups
in volume IV, Basic Analysis of Regularized Series and Products (with Jay
Jorgenson in volume V, and Explicit Formulas for Regularized Products and Series
(with Jay Jorgenson) in volume V • THE FILE • CHALLENGES

Serge Lang

Introduction to Differentiable Manifolds

Second Edition

With 12 Illustrations

Springer

Serge Lang
Department of Mathematics
Yale University
New Haven, CT 06520
USA

Editorial Board
(North America):

S. Axler
Mathematics Department
San Francisco State University
San Francisco, CA 94132
USA
axler@sfsu.edu

F.W. Gehring
Mathematics Department
East Hall
University of Michigan
Ann Arbor, MI 48109-1109
USA
fgehring@math.lsa.umich.edu

K.A. Ribet
Mathematics Department
University of California, Berkeley
Berkeley, CA 94720-3840
USA
ribet@math.berkeley.edu

Mathematics Subject Classification (2000): 58Axx, 34M45, 57Nxx, 57Rxx

Library of Congress Cataloging-in-Publication Data
Lang, Serge, 1927–
    Introduction to differentiable manifolds / Serge Lang. — 2nd ed.
      p.   cm. — (Universitext)
    Includes bibliographical references and index.
    ISBN 978-1-4419-3019-4   ISBN 978-0-387-21772-7 (eBook)
    DOI 10.1007/978-0-387-21772-7
    1. Differential topology. 2. Differentiable manifolds. I. Title.
  QA649.L3   2002
  516.3´6—dc21   2002020940

The first edition of this book was published by Addison-Wesley, Reading, MA, 1972.

ISBN 978-1-4419-3019-4   Printed on acid-free paper.

© 2002 Springer Science+Business Media New York, corrected publication 2021
Originally published by Springer-Verlag New York, Inc. in 2002
Softcover reprint of the hardcover 1st edition 2002

9 8 7 6 5 4 3 2 1   SPIN 10874516

www.springer-ny.com

Foreword

This book is an outgrowth of my *Introduction to Differentiable Manifolds* (1962) and *Differential Manifolds* (1972). Both I and my publishers felt it worth while to keep available a brief introduction to differential manifolds.

The book gives an introduction to the basic concepts which are used in differential topology, differential geometry, and differential equations. In differential topology, one studies for instance homotopy classes of maps and the possibility of finding suitable differentiable maps in them (immersions, embeddings, isomorphisms, etc.). One may also use differentiable structures on topological manifolds to determine the topological structure of the manifold (for example, à la Smale [Sm 67]). In differential geometry, one puts an additional structure on the differentiable manifold (a vector field, a spray, a 2-form, a Riemannian metric, ad lib.) and studies properties connected especially with these objects. Formally, one may say that one studies properties invariant under the group of differentiable automorphisms which preserve the additional structure. In differential equations, one studies vector fields and their integral curves, singular points, stable and unstable manifolds, etc. A certain number of concepts are essential for all three, and are so basic and elementary that it is worthwhile to collect them together so that more advanced expositions can be given without having to start from the very beginnings. The concepts are concerned with the general basic theory of differential manifolds. My *Fundamentals of Differential Geometry* (1999) can then be viewed as a continuation of the present book.

Charts and local coordinates. A chart on a manifold is classically a representation of an open set of the manifold in some euclidean space. Using a chart does not necessarily imply using coordinates. Charts will be used systematically.

I don't propose, of course, to do away with local coordinates. They are useful for computations, and are also especially useful when integrating differential forms, because the $dx_1 \wedge \cdots \wedge dx_n$. corresponds to the $dx_1 \cdots dx_n$ of Lebesgue measure, in oriented charts. Thus we often give the local coordinate formulation for such applications. Much of the literature is still covered by local coordinates, and I therefore hope that the neophyte will thus be helped in getting acquainted with the literature. I also hope to convince the expert that nothing is lost, and much is gained, by expressing one's geometric thoughts without hiding them under an irrelevant formalism.

Since this book is intended as a text to follow advanced calculus, say at the first year graduate level or advanced undergraduate level, manifolds are assumed finite dimensional. Since my book *Fundamentals of Differential Geometry* now exists, and covers the infinite dimensional case as well, readers at a more advanced level can verify for themselves that there is no essential additional cost in this larger context. I am, however, following here my own admonition in the introduction of that book, to assume from the start that all manifolds are finite dimensional. Both presentations need to be available, for mathematical and pedagogical reasons.

New Haven 2002   Serge Lang

The original version of this book was revised: The title of this book was incorrectly captured as Introduction to Differential Manifolds. The correct title should be Introduction to Differentiable Manifolds.This has been corrected. A correction of this book can be found at DOI 10.1007/0-387-21772-X_11.

Acknowledgments

I have greatly profited from several sources in writing this book. These sources are from the 1960s.

First, I originally profited from Dieudonné's *Foundations of Modern Analysis*, which started to emphasize the Banach point of view.

Second, I originally profited from Bourbaki's *Fascicule de résultats* [Bou 69] for the foundations of differentiable manifolds. This provides a good guide as to what should be included. I have not followed it entirely, as I have omitted some topics and added others, but on the whole, I found it quite useful. I have put the emphasis on the differentiable point of view, as distinguished from the analytic. However, to offset this a little, I included two analytic applications of Stokes' formula, the Cauchy theorem in several variables, and the residue theorem.

Third, Milnor's notes [Mi 58], [Mi 59], [Mi 61] proved invaluable. They were of course directed toward differential topology, but of necessity had to cover ad hoc the foundations of differentiable manifolds (or, at least, part of them). In particular, I have used his treatment of the operations on vector bundles (Chapter III, §4) and his elegant exposition of the uniqueness of tubular neighborhoods (Chapter IV, §6, and Chapter VII, §4).

Fourth, I am very much indebted to Palais for collaborating on Chapter IV, and giving me his exposition of sprays (Chapter IV, §3). As he showed me, these can be used to construct tubular neighborhoods. Palais also showed me how one can recover sprays and geodesics on a Riemannian manifold by making direct use of the canonical 2-form and the metric (Chapter VII, §7). This is a considerable improvement on past expositions.

Contents

CHAPTER I

Differential Calculus

We shall recall briefly the notion of derivative and some of its useful properties. My books on analysis [La83/97], [La 93] give a self-contained and complete treatment. We summarize basic facts of the differential calculus. *The reader can actually skip this chapter* and start immediately with Chapter II if the reader is accustomed to thinking about the derivative of a map as a linear transformation. (In the finite dimensional case, when bases have been selected, the entries in the matrix of this transformation are the partial derivatives of the map.) We have repeated the proofs for the more important theorems, for the ease of the reader.

It is convenient to use throughout the language of categories. The notion of category and morphism (whose definitions we recall in §1) is designed to abstract what is common to certain collections of objects and maps between them. For instance, euclidean vector spaces and linear maps, open subsets of euclidean spaces and differentiable maps, differentiable manifolds and differentiable maps, vector bundles and vector bundle maps, topological spaces and continuous maps, sets and just plain maps. In an arbitrary category, maps are called morphisms, and in fact the category of differentiable manifolds is of such importance in this book that from Chapter II on, we use the word morphism synonymously with differentiable map (or p-times differentiable map, to be precise). All other morphisms in other categories will be qualified by a prefix to indicate the category to which they belong.

I, §1. CATEGORIES

A **category** is a collection of objects $\{X, Y, \ldots\}$ such that for two objects X, Y we have a set $\text{Mor}(X, Y)$ and for three objects X, Y, Z a mapping (composition law)

$$\text{Mor}(X, Y) \times \text{Mor}(Y, Z) \to \text{Mor}(X, Z)$$

satisfying the following axioms:

CAT 1. *Two sets $\text{Mor}(X, Y)$ and $\text{Mor}(X', Y')$ are disjoint unless $X = X'$ and $Y = Y'$, in which case they are equal.*

CAT 2. *Each $\text{Mor}(X, X)$ has an element id_X which acts as a left and right identity under the composition law.*

CAT 3. *The composition law is associative.*

The elements of $\text{Mor}(X, Y)$ are called **morphisms**, and we write frequently $f \colon X \to Y$ for such a morphism. The composition of two morphisms f, g is written fg or $f \circ g$.

Elements of $\text{Mor}(X, X)$ are called **endomorphisms** of X, and we write

$$\text{Mor}(X, X) = \text{End}(X).$$

For a more extensive description of basic facts about categories, see my *Algebra* [La 02], Chapter I, §1. Here we just remind the reader of the basic terminology which we use. The main categories for us will be:

Vector spaces, whose morphisms are linear maps.

Open sets in a finite dimensional vector space over **R**, whose morphisms are differentiable maps (of given degree of differentiability, $C^0, C^1, \ldots,$ C^∞).

Manifolds, with morphisms corresponding to the morphisms just mentioned. See Chapter II, §1.

In any category, a morphism $f \colon X \to Y$ is said to be an **isomorphism** if it has an inverse in the category, that is, there exists a morphism $g \colon Y \to X$ such that fg and gf are the identities (of Y and X respectively). An isomorphism in the category of topological spaces (whose morphisms are continuous maps) has been called a **homeomorphism**. We stick to the functorial language, and call it a **topological isomorphism**. In general, we describe the category to which a morphism belongs by a suitable prefix. In the category of sets, a set-isomorphism is also called a **bijection**. **Warning**: A map $f \colon X \to Y$ may be an isomorphism in one category but not in another. For example, the map $x \mapsto x^3$ from $\mathbf{R} \to \mathbf{R}$ is a C^0-isomorphism, but not a C^1 isomorphism (the inverse is continuous, but not differentiable at the origin). In the category of vector spaces, it is true that a bijective morphism is an isomorphism, but the example we just gave shows that the conclusion does not necessarily hold in other categories.

An **automorphism** is an isomorphism of an object with itself. The set of automorphisms of an object X in a category form a group, denoted by $\mathrm{Aut}(X)$.

If $f: X \to Y$ is a morphism, then a **section** of f is defined to be a morphism $g: Y \to X$ such that $f \circ g = \mathrm{id}_Y$.

A **functor** $\lambda: \mathfrak{A} \to \mathfrak{A}'$ from a category $\mathfrak{A}$ into a category $\mathfrak{A}'$ is a map which associates with each object X in $\mathfrak{A}$ an object $\lambda(X)$ in $\mathfrak{A}'$, and with each morphism $f: X \to Y$ a morphism $\lambda(f): \lambda(X) \to \lambda(Y)$ in $\mathfrak{A}'$ such that, whenever f and g are morphisms in $\mathfrak{A}$ which can be composed, then $\lambda(fg) = \lambda(f)\lambda(g)$ and $\lambda(\mathrm{id}_X) = \mathrm{id}_{\lambda(X)}$ for all X. This is in fact a covariant functor, and a contravariant functor is defined by reversing the arrows (so that we have $\lambda(f): \lambda(Y) \to \lambda(X)$ and $\lambda(fg) = \lambda(g)\lambda(f)$).

In a similar way, one defines functors of many variables, which may be covariant in some variables and contravariant in others. We shall meet such functors when we discuss multilinear maps, differential forms, etc.

The functors of the same variance from one category $\mathfrak{A}$ to another $\mathfrak{A}'$ form themselves the objects of a category $\mathrm{Fun}(\mathfrak{A}, \mathfrak{A}')$. Its morphisms will sometimes be called **natural transformations** instead of functor morphisms. They are defined as follows. If λ, μ are two functors from $\mathfrak{A}$ to $\mathfrak{A}'$ (say covariant), then a natural transformation $t: \lambda \to \mu$ consists of a collection of morphisms

$$t_X: \lambda(X) \to \mu(X)$$

as X ranges over $\mathfrak{A}$, which makes the following diagram commutative for any morphism $f: X \to Y$ in $\mathfrak{A}$:

$$
\begin{array}{ccc}
\lambda(X) & \xrightarrow{\ t_X\ } & \mu(X) \\
\lambda(f) \downarrow & & \downarrow \mu(f) \\
\lambda(Y) & \xrightarrow[\ t_Y\ ]{} & \mu(Y)
\end{array}
$$

Vector spaces form a category, the morphisms being the linear maps. Note that $(E, F) \mapsto L(E, F)$ is a functor in two variables, contravariant in the first variable and covariant in the second. If many categories are being considered simultaneously, then an isomorphism in the category of vector spaces and linear map is called a **linear** isomorphism. We write $\mathrm{Lis}(E, F)$ and $\mathrm{Laut}(E)$ for the vector spaces of linear isomorphisms of E onto F, and the linear automorphisms of E respectively.

The vector space of r-multilinear maps

$$\psi: E \times \cdots \times E \to F$$

of E into F will be denoted by $L^r(E, F)$. Those which are symmetric (resp. alternating) will be denoted by $L^r_s(E, F)$ or $L^r_{\text{sym}}(E, F)$ (resp. $L^r_a(E, F)$). **Symmetric** means that the map is invariant under a permutation of its variables. **Alternating** means that under a permutation, the map changes by the sign of the permutation.

We find it convenient to denote by $L(\mathbf{E})$, $L^r(\mathbf{E})$, $L^r_s(\mathbf{E})$, and $L^r_a(\mathbf{E})$ the linear maps of $\mathbf{E}$ into $\mathbf{R}$ (resp. the r-multilinear, symmetric, alternating maps of $\mathbf{E}$ into $\mathbf{R}$). Following classical terminology, it is also convenient to call such maps into $\mathbf{R}$ **forms** (of the corresponding type). If $\mathbf{E}_1, \dots, \mathbf{E}_r$ and $\mathbf{F}$ are vector spaces, then we denote by $L(\mathbf{E}_1, \dots, \mathbf{E}_r; \mathbf{F})$ the multilinear maps of the product $\mathbf{E}_1 \times \cdots \times \mathbf{E}_r$ into $\mathbf{F}$. We let:

$$\text{End}(\mathbf{E}) = L(\mathbf{E}, \mathbf{E}),$$

$\text{Laut}(\mathbf{E}) = $ elements of $\text{End}(\mathbf{E})$ which are invertible in $\text{End}(\mathbf{E})$.

Thus for our finite dimensional vector space E, an element of $\text{End}(\mathbf{E})$ is in $\text{Laut}(\mathbf{E})$ if and only if its determinant is $\neq 0$.

Suppose E, F are given norms. They determine a natural norm on $L(E, F)$, namely for $A \in L(E, F)$, the **operator norm** $|A|$ is the greatest lower bound of all numbers K such that

$$|Ax| \leq K|x|$$

for all $x \in \mathbf{E}$.

I, §2. FINITE DIMENSIONAL VECTOR SPACES

Unless otherwise specified, **vector spaces** *will be finite dimensional over the real numbers.* Such vector spaces are linearly isomorphic to euclidean space $\mathbf{R}^n$ for some n. They have norms. If a basis $\{e_1, \dots, e_n\}$ is selected, then there are two natural norms: the euclidean norm, such that for a vector v with coordinates $(x_1, \dots, x_n)$ with respect to the basis, we have

$$|v|^2_{\text{euc}} = x_1^2 + \cdots + x_n^2.$$

The other natural norm is the sup norm, written $|v|_\infty$, such that

$$|v|_\infty = \max_i |x_i|.$$

It is an elementary lemma that all norms on a finite dimensional vector space $\mathbf{E}$ are equivalent. In other words, if $| \ |_1$ and $| \ |_2$ are norms on $\mathbf{E}$, then there exist constants $C_1, C_2 > 0$ such that for all $v \in \mathbf{E}$ we have

$$C_1|v|_1 \leq |v|_2 \leq C_2|v|_1.$$

A vector space with a norm is called a **normed vector space**. They form a category whose morphisms are the norm preserving linear maps, which are then necessarily injective.

By a **euclidean space** we mean a vector space with a positive definite scalar product. A morphism in the euclidean category is a linear map which preserves the scalar product. Such a map is necessarily injective. An isomorphism in this category is called a **metric** or **euclidean isomorphism**. An orthonormal basis of a euclidean vector space gives rise to a metric isomorphism with $\mathbf{R}^n$, mapping the unit vectors in the basis on the usual unit vectors of $\mathbf{R}^n$.

Let $\mathbf{E}$, $\mathbf{F}$ be vector spaces (so finite dimensional over $\mathbf{R}$ by convention). The set of linear maps from $\mathbf{E}$ into $\mathbf{F}$ is a vector space isomorphic to the space of $m \times n$ matrices if dim $\mathbf{E} = m$ and dim $\mathbf{F} = n$.

Note that $(\mathbf{E}, \mathbf{F}) \mapsto L(\mathbf{E}, \mathbf{F})$ is a functor, contravariant in $\mathbf{E}$ and covariant in $\mathbf{F}$. Similarly, we have the vector space of multilinear maps

$$L(\mathbf{E}_1, \ldots, \mathbf{E}_r, \mathbf{F})$$

of a product $\mathbf{E}_1 \times \cdots \times \mathbf{E}_r$ into $\mathbf{F}$. Suppose norms are given on all $\mathbf{E}_i$ and $\mathbf{F}$. Then a natural norm can be defined on $L(\mathbf{E}_1, \ldots, \mathbf{E}_r, \mathbf{F})$, namely the norm of a multilinear map

$$A: \mathbf{E}_1 \times \cdots \times \mathbf{E}_r \to \mathbf{F}$$

is defined to be the greatest lower bound of all numbers K such that

$$|A(x_1, \ldots, x_r)| \leqq K|x_1| \cdots |x_r|.$$

We have:

Proposition 2.1. *The canonical map*

$$L\big(\mathbf{E}_1, L(\mathbf{E}_2, \ldots, L(\mathbf{E}_r, \mathbf{F}))\big) \to L'(\mathbf{E}_1, \ldots, \mathbf{E}_r, \mathbf{F})$$

from the repeated linear maps to the multilinear maps is a linear isomorphism which is norm preserving.

For purely differential properties, which norms are chosen are irrelevant since all norms are equivalent. The relevance will arise when we deal with metric structures, called Riemannian, in Chapter VII.

We note that a linear map and a multilinear map are necessarily continuous, having assumed the vector spaces to be finite dimensional.

I, §3. DERIVATIVES AND COMPOSITION OF MAPS

For the calculus in vector spaces, see my *Undergraduate Analysis* [La 83/97]. We recall some of the statements here.

A real valued function of a real variable, defined on some neighborhood of 0 is said to be $o(t)$ if

$$\lim_{t \to 0} o(t)/t = 0.$$

Let **E**, **F** be two vector spaces (assumed finite dimensional), and φ a mapping of a neighborhood of 0 in **E** into **F**. We say that φ is **tangent to 0** if, given a neighborhood W of 0 in **F**, there exists a neighborhood V of 0 in **E** such that

$$\varphi(tV) \subset o(t)W$$

for some function $o(t)$. If both **E**, **F** are normed, then this amounts to the usual condition

$$|\varphi(x)| \leq |x|\psi(x)$$

with $\lim \psi(x) = 0$ as $|x| \to 0$.

Let **E**, **F** be two vector spaces and U open in **E**. Let $f: U \to \mathbf{F}$ be a continuous map. We shall say that f is **differentiable** at a point $x_0 \in U$ if there exists a linear map λ of **E** into **F** such that, if we let

$$f(x_0 + y) = f(x_0) + \lambda y + \varphi(y)$$

for small y, then φ is tangent to 0. It then follows trivially that λ is uniquely determined, and we say that it is the **derivative** of f at x_0. We denote the derivative by $Df(x_0)$ or $f'(x_0)$. It is an element of $L(\mathbf{E}, \mathbf{F})$. If f is differentiable at every point of U, then f' is a map

$$f': U \to L(\mathbf{E}, \mathbf{F}).$$

It is easy to verify the chain rule.

Proposition 3.1. *If $f: U \to V$ is differentiable at x_0, if $g: V \to W$ is differentiable at $f(x_0)$, then $g \circ f$ is differentiable at x_0, and*

$$(g \circ f)'(x_0) = g'\big(f(x_0)\big) \circ f'(x_0).$$

Proof. We leave it as a simple (and classical) exercise.

The rest of this section is devoted to the statements of the differential calculus.

Let U be open in **E** and let $f: U \to \mathbf{F}$ be differentiable at each point of U. If f' is continuous, then we say that f is **of class** C^1. We define maps of class C^p ($p \geqq 1$) inductively. The p-th derivative $D^p f$ is defined as $D(D^{p-1} f)$ and is itself a map of U into

$$L\big(\mathbf{E}, \, L(\mathbf{E}, \ldots, L(\mathbf{E}, \mathbf{F}))\big)$$

which can be identified with $L^p(\mathbf{E}, \mathbf{F})$ by Proposition 2.1. A map f is said to be **of class** C^p if its kth derivative $D^k f$ exists for $1 \leqq k \leqq p$, and is continuous.

Remark. *Let f be of class C^p, on an open set U containing the origin. Suppose that f is locally homogeneous of degree p near 0, that is*

$$f(tx) = t^p f(x)$$

for all t and x sufficiently small. Then for all sufficiently small x we have

$$f(x) = \frac{1}{p!} D^p f(0) x^{(p)},$$

where $x^{(p)} = (x, x, \ldots, x)$, p times.

This is easily seen by differentiating p times the two expressions for $f(tx)$, and then setting $t = 0$. The differentiation is a trivial application of the chain rule.

Proposition 3.2. *Let U, V be open in vector spaces. If $f: U \to V$ and $g: V \to \mathbf{F}$ are of class C^p, then so is $g \circ f$.*

From Proposition 3.2, we can view open subsets of vector spaces as the objects of a category, whose morphisms are the continuous maps of class C^p. These will be called C^p-**morphisms**. We say that f is of class C^∞ if it is of class C^p for all integers $p \geqq 1$. From now on, p is an integer $\geqq 0$ or ∞ (C^0 maps being the continuous maps). In practice, we omit the prefix C^p if the p remains fixed. Thus by **morphism**, throughout the rest of this book, we mean C^p-morphism with $p \leqq \infty$. We shall use the word morphism also for C^p-morphisms of manifolds (to be defined in the next chapter), *but morphisms in any other category will always be prefixed so as to indicate the category to which they belong* (for instance bundle morphism, continuous linear morphism, etc.).

Proposition 3.3. *Let U be open in the vector space $\mathbf{E}$, and let $f: U \to \mathbf{F}$ be a C^p-morphism. Then $D^p f$ (viewed as an element of $L^p(\mathbf{E}, \mathbf{F})$) is symmetric.*

Proposition 3.4. *Let U be open in $\mathbf{E}$, and let $f_i: U \to \mathbf{F}_i$ ($i = 1, \ldots, n$) be continuous maps into spaces $\mathbf{F}_i$. Let $f = (f_1, \ldots, f_n)$ be the map of U*

into the product of the $\mathbf{F}_i$*. Then* f *is of class* C^p *if and only if each* f_i *is of class* C^p*, and in that case*

$$D^p f = (D^p f_1, \ldots, D^p f_n).$$

Let U, V be open in spaces $\mathbf{E}_1$, $\mathbf{E}_2$ and let

$$f \colon U \times V \to \mathbf{F}$$

be a continuous map into a vector space. We can introduce the notion of partial derivative in the usual manner. If (x, y) is in $U \times V$ and we keep y fixed, then as a function of the first variable, we have the derivative as defined previously. This derivative will be denoted by $D_1 f(x, y)$. Thus

$$D_1 f \colon U \times V \to L(\mathbf{E}_1, \mathbf{F})$$

is a map of $U \times V$ into $L(\mathbf{E}_1, \mathbf{F})$. We call it the **partial derivative** with respect to the first variable. Similarly, we have $D_2 f$, and we could take n factors instead of 2. The total derivative and the partials are then related as follows.

Proposition 3.5. *Let* $U_1, \ldots, U_n$ *be open in the spaces* $\mathbf{E}_1, \ldots, \mathbf{E}_n$ *and let* $f \colon U_1 \times \cdots \times U_n \to \mathbf{F}$ *be a continuous map. Then* f *is of class* C^p *if and only if each partial derivative* $D_i f \colon U_1 \times \cdots U_n \to L(\mathbf{E}_i, \mathbf{F})$ *exists and is of class* C^{p-1}*. If that is the case, then for* $x = (x_1, \ldots, x_n)$ *and*

$$v = (v_1, \ldots, v_n) \in \mathbf{E}_1 \times \cdots \times \mathbf{E}_n,$$

we have

$$Df(x) \cdot (v_1, \ldots, v_n) = \sum D_i f(x) \cdot v_i.$$

The next four propositions are concerned with continuous linear and multilinear maps.

Proposition 3.6. *Let* $\mathbf{E}$*,* $\mathbf{F}$ *be vector spaces and* $f \colon \mathbf{E} \to \mathbf{F}$ *a continuous linear map. Then for each* $x \in \mathbf{E}$ *we have*

$$f'(x) = f.$$

Proposition 3.7. *Let* $\mathbf{E}$*,* $\mathbf{F}$*,* $\mathbf{G}$ *be vector spaces, and* U *open in* $\mathbf{E}$*. Let* $f \colon U \to \mathbf{F}$ *be of class* C^p *and* $g \colon \mathbf{F} \to \mathbf{G}$ *linear. Then* $g \circ f$ *is of class* C^p *and*

$$D^p(g \circ f) = g \circ D^p f.$$

Proposition 3.8. *If* $\mathbf{E}_1, \ldots, \mathbf{E}_r$ *and* $\mathbf{F}$ *are vector spaces and*

$$f \colon \mathbf{E}_1 \times \cdots \times \mathbf{E}_r \to \mathbf{F}$$

a multilinear map, then f is of class C^∞, and its $(r+1)$-st derivative is 0. If $r = 2$, then Df is computed according to the usual rule for derivative of a product (*first times the derivative of the second plus derivative of the first times the second*).

Proposition 3.9. *Let* **E**, **F** *be vector spaces which are isomorphic. If* $u: \mathbf{E} \to \mathbf{F}$ *is an isomorphism, we denote its inverse by* u^{-1}. *Then the map*

$$u \mapsto u^{-1}$$

from Lis(**E**, **F**) *to* Lis(**F**, **E**) *is a* C^∞-*isomorphism. Its derivative at a point* u_0 *is the linear map of* $L(\mathbf{E}, \mathbf{F})$ *into* $L(\mathbf{F}, \mathbf{E})$ *given by the formula*

$$v \mapsto u_0^{-1} v u_0^{-1}.$$

Finally, we come to some statements which are of use in the theory of vector bundles.

Proposition 3.10. *Let* U *be open in the vector space* **E** *and let* **F**, **G** *be vector spaces.*

(i) *If* $f: U \to L(\mathbf{E}, \mathbf{F})$ *is a* C^p-*morphism, then the map of* $U \times \mathbf{E}$ *into* **F** *given by*
$$(x, v) \mapsto f(x)v$$
is a morphism.

(ii) *If* $f: U \to L(\mathbf{E}, \mathbf{F})$ *and* $g: U \to L(\mathbf{F}, \mathbf{G})$ *are morphisms, then so is* $\gamma(f, g)$ (γ *being the composition*).

(iii) *If* $f: U \to \mathbf{R}$ *and* $g: U \to L(\mathbf{E}, \mathbf{F})$ *are morphisms, so is* fg (*the value of* fg *at* x *is* $f(x)g(x)$, *ordinary multiplication by scalars*).

(iv) *If* $f, g: U \to L(\mathbf{E}, \mathbf{F})$ *are morphisms, so is* $f + g$.

This proposition concludes our summary of results assumed without proof.

I, §4. INTEGRATION AND TAYLOR'S FORMULA

Let **E** be a vector space. We continue to assume finite dimensionality over **R**. Let I denote a real, closed interval, say $a \le t \le b$. A **step mapping**

$$f: I \to \mathbf{E}$$

is a mapping such that there exists a finite number of disjoint sub-intervals $I_1, \ldots, I_n$ covering I such that on each interval I_j, the mapping has constant value, say v_j. We do not require the intervals I_j to be closed. They may be open, closed, or half-closed.

Given a sequence of mappings f_n from I into $\mathbf{E}$, we say that it converges uniformly if, given a neighborhood W of 0 into $\mathbf{E}$, there exists an integer n_0 such that, for all n, $m > n_0$ and all $t \in I$, the difference $f_n(t) - f_m(t)$ lies in W. The sequence f_n then converges to a mapping f of I into $\mathbf{E}$.

A **ruled** mapping is a uniform limit of step mappings. We leave to the reader the proof that every continuous mapping is ruled.

If f is a step mapping as above, we define its integral

$$\int_a^b f = \int_a^b f(t)\, dt = \sum \mu(I_j) v_j,$$

where $\mu(I_j)$ is the length of the interval I_j (its measure in the standard Lebesgue measure). This integral is independent of the choice of intervals I_j on which f is constant.

If f is ruled and $f = \lim f_n$ (lim being the uniform limit), then the sequence

$$\int_a^b f_n$$

converges in $\mathbf{E}$ to an element of $\mathbf{E}$ independent of the particular sequence f_n used to approach f uniformly. We denote this limit by

$$\int_a^b f = \int_a^b f(t)\, dt$$

and call it the **integral** of f. The integral is linear in f, and satisfies the usual rules concerning changes of intervals. (If $b < a$ then we define $\int_a^b$ to be minus the integral from b to a.)

As an immediate consequence of the definition, we get:

Proposition 4.1. *Let* $\lambda \colon \mathbf{E} \to \mathbf{R}$ *be a linear map and let* $f \colon I \to \mathbf{E}$ *be ruled. Then* $\lambda f = \lambda \circ f$ *is ruled, and*

$$\lambda \int_a^b f(t)\, dt = \int_a^b \lambda f(t)\, dt.$$

Proof. If f_n is a sequence of step functions converging uniformly to f, then λf_n is ruled and converges uniformly to λf. Our formula follows at once.

Taylor's Formula. *Let* $\mathbf{E}$, $\mathbf{F}$ *be vector spaces. Let* U *be open in* $\mathbf{E}$. *Let* x, y *be two points of* U *such that the segment* $x + ty$ *lies in* U *for* $0 \leq t \leq 1$. *Let*

$$f \colon U \to \mathbf{F}$$

be a C^p-morphism, and denote by $y^{(p)}$ the "vector" $(y, \ldots, y)$ p times. Then the function $D^p f(x + ty) \cdot y^{(p)}$ is continuous in t, and we have

$$f(x + y) = f(x) + \frac{Df(x)y}{1!} + \cdots + \frac{D^{p-1}f(x)y^{(p-1)}}{(p-1)!}$$
$$+ \int_0^1 \frac{(1 - t)^{p-1}}{(p-1)!} D^p f(x + ty) y^{(p)} \, dt.$$

Proof. It suffices to show that both sides give the same thing when we apply a functional λ (linear map into $\mathbf{R}$). This follows at once from Proposition 3.7 and 4.1, together with the known result when $\mathbf{F} = \mathbf{R}$. In this case, the proof proceeds by induction on p, and integration by parts, starting from

$$f(x + y) - f(x) = \int_0^1 Df(x + ty)y \, dt.$$

The next two corollaries are known as the **mean value theorem**.

Corollary 4.2. *Let $\mathbf{E}$, $\mathbf{F}$ be two normed vector spaces, U open in $\mathbf{E}$. Let x, z be two distinct points of U such that the segment $x + t(z - x)$ $(0 \leq t \leq 1)$ lies in U. Let $f \colon U \to \mathbf{F}$ be continuous and of class C^1. Then*

$$|f(z) - f(x)| \leq |z - x| \sup |f'(\xi)|,$$

the sup being taken over ξ in the segment.

Proof. This comes from the usual estimations of the integral. Indeed, for any continuous map $g \colon I \to \mathbf{F}$ we have the estimate

$$\left| \int_a^b g(t) \, dt \right| \leq K(b - a)$$

if K is a bound for g on I, and $a \leq b$. This estimate is obvious for step functions, and therefore follows at once for continuous functions.

Another version of the mean value theorem is frequently used.

Corollary 4.3. *Let the hypotheses be as in Corollary 4.2. Let x_0 be a point on the segment between x and z. Then*

$$|f(z) - f(x) - f'(x_0)(z - x)| \leq |z - x| \sup |f'(\xi) - f'(x_0)|,$$

the sup taken over all ξ on the segment.

Proof. We apply Corollary 4.2 to the map

$$g(x) = f(x) - f'(x_0)x.$$

Finally, let us make some comments on the estimate of the remainder term in Taylor's formula. We have assumed that $D^p f$ is continuous. Therefore, $D^p f(x + ty)$ can be written

$$D^p f(x + ty) = D^p f(x) + \psi(y, t),$$

where ψ depends on y, t (and x of course), and for fixed x, we have

$$\lim |\psi(y, t)| = 0$$

as $|y| \to 0$. Thus we obtain:

Corollary 4.4. *Let* **E**, **F** *be two normed vector spaces, U open in* **E**, *and x a point of U. Let* $f : U \to \mathbf{F}$ *be of class* C^p, $p \geq 1$. *Then for all y such that the segment* $x + ty$ *lies in U* $(0 \leq t \leq 1)$, *we have*

$$f(x + y) = f(x) + \frac{Df(x)y}{1!} + \cdots + \frac{D^p f(x) y^{(p)}}{p!} + \theta(y)$$

with an error term $\theta(y)$ *satisfying*

$$\lim_{y \to 0} \theta(y)/|y|^p = 0.$$

I, §5. THE INVERSE MAPPING THEOREM

The inverse function theorem and the existence theorem for differential equations (of Chapter IV) are based on the next result.

Lemma 5.1 (Contraction Lemma or Shrinking Lemma). *Let M be a complete metric space, with distance function d, and let* $f : M \to M$ *be a mapping of M into itself. Assume that there is a constant K, $0 < K < 1$, such that, for any two points x, y in M, we have*

$$d(f(x), f(y)) \leq K \, d(x, y).$$

Then f has a unique fixed point (a point such that $f(x) = x$). *Given any point* x_0 *in M, the fixed point is equal to the limit of* $f^n(x_0)$ *(iteration of f repeated n times) as n tends to infinity.*

Proof. This is a trivial exercise in the convergence of the geometric series, which we leave to the reader.

Theorem 5.2. *Let* $\mathbf{E}$, $\mathbf{F}$ *be normed vector spaces,* U *an open subset of* $\mathbf{E}$, *and let* $f: U \to \mathbf{F}$ *a* C^p-*morphism with* $p \geq 1$. *Assume that for some point* $x_0 \in U$, *the derivative* $f'(x_0): \mathbf{E} \to \mathbf{F}$ *is a linear isomorphism. Then* f *is a local* C^p-*isomorphism at* x_0.

(By a **local** C^p-**isomorphism** at x_0, we mean that there exists an open neighborhood V of x_0 such that the restriction of f to V establishes a C^p-isomorphism between V and an open subset of $\mathbf{E}$.)

Proof. Since a linear isomorphism is a C^∞-isomorphism, we may assume without loss of generality that $\mathbf{E} = \mathbf{F}$ and $f'(x_0)$ is the identity (simply by considering $f'(x_0)^{-1} \circ f$ instead of f). After translations, we may also assume that $x_0 = 0$ and $f(x_0) = 0$.

We let $g(x) = x - f(x)$. Then $g'(x_0) = 0$ and by continuity there exists $r > 0$ such that, if $|x| < 2r$, we have

$$|g'(x)| < \tfrac{1}{2}.$$

From the mean value theorem, we see that $|g(x)| \leq \tfrac{1}{2}|x|$ and hence g maps the closed ball of radius r, $\bar{B}_r(0)$, into $\bar{B}_{r/2}(0)$.

We contend: Given $y \in \bar{B}_{r/2}(0)$, there exists a unique element $x \in \bar{B}_r(0)$ such that $f(x) = y$. We prove this by considering the map

$$g_y(x) = y + x - f(x).$$

If $|y| \leq r/2$ and $|x| \leq r$, then $|g_y(x)| \leq r$ and hence g_y may be viewed as a mapping of the complete metric space $\bar{B}_r(0)$ into itself. The bound of $\tfrac{1}{2}$ on the derivative together with the mean value theorem shows that g_y is a contracting map, i.e. that

$$|g_y(x_1) - g_y(x_2)| = |g(x_1) - g(x_2)| \leq \tfrac{1}{2}|x_1 - x_2|$$

for x_1, $x_2 \in \bar{B}_r(0)$. By the contraction lemma, it follows that g_y has a unique fixed point. But the fixed point of g_y is precisely the solution of the equation $f(x) = y$. This proves our contention.

We obtain a local inverse $\varphi = f^{-1}$. This inverse is continuous, because

$$|x_1 - x_2| \leq |f(x_1) - f(x_2)| + |g(x_1) - g(x_2)|$$

and hence

$$|x_1 - x_2| \leq 2|f(x_1) - f(x_2)|.$$

Furthermore φ is differentiable in $B_{r/2}(0)$. Indeed, let $y_1 = f(x_1)$ and $y_2 = f(x_2)$ with $y_1, y_2 \in B_{r/2}(0)$ and $x_1, x_2 \in \bar{B}_r(0)$. Then

$$|\varphi(y_1) - \varphi(y_2) - f'(x_2)^{-1}(y_1 - y_2)| = |x_1 - x_2 - f'(x_2)^{-1}(f(x_1) - f(x_2))|.$$

We operate on the expression inside the norm sign with the identity

$$\mathrm{id} = f'(x_2)^{-1} f'(x_2).$$

Estimating and using the continuity of f', we see that for some constant A, the preceding expression is bounded by

$$A|f'(x_2)(x_1 - x_2) - f(x_1) + f(x_2)|.$$

From the differentiability of f, we conclude that this expression is $o(x_1 - x_2)$ which is also $o(y_1 - y_2)$ in view of the continuity of φ proved above. This proves that φ is differentiable and also that its derivative is what it should be, namely

$$\varphi'(y) = f'(\varphi(y))^{-1},$$

for $y \in B_{r/2}(0)$. Since the mappings φ, f', "inverse" are continuous, it follows that φ' is continuous and thus that φ is of class C^1. Since taking inverses is C^∞ and f' is C^{p-1}, it follows inductively that φ is C^p, as was to be shown.

Note that this last argument also proves:

Proposition 5.3. *If $f: U \to V$ is a homeomorphism and is of class C^p with $p \geq 1$, and if f is a C^1-isomorphism, then f is a C^p-isomorphism.*

In some applications it is necessary to know that if the derivative of a map is close to the identity, then the image of a ball contains a ball of only slightly smaller radius. The precise statement follows. In this book, it will be used only in the proof of the change of variables formula, and therefore may be omitted until the reader needs it.

Lemma 5.4. *Let U be open in $\mathbf{E}$, and let $f: U \to \mathbf{E}$ be of class C^1. Assume that $f(0) = 0$, $f'(0) = I$. Let $r > 0$ and assume that $\bar{B}_r(0) \subset U$. Let $0 < s < 1$, and assume that*

$$|f'(z) - f'(x)| \leq s$$

for all $x, z \in \bar{B}_r(0)$. If $y \in \mathbf{E}$ and $|y| \leq (1-s)r$, then there exists a unique $x \in \bar{B}_r(0)$ such that $f(x) = y$.

Proof. The map g_y given by $g_y(x) = x - f(x) + y$ is defined for $|x| \leq r$ and $|y| \leq (1 - s)r$, and maps $\bar{B}_r(0)$ into itself because, from the estimate

$$|f(x) - x| = |f(x) - f(0) - f'(0)x| \leq |x| \sup |f'(z) - f'(0)| \leq sr,$$

we obtain

$$|g_y(x)| \leq sr + (1 - s)r = r.$$

Furthermore, g_y is a shrinking map because, from the mean value theorem, we get

$$\begin{aligned}
|g_y(x_1) - g_y(x_2)| &= |x_1 - x_2 - (f(x_1) - f(x_2))| \\
&= |x_1 - x_2 - f'(0)(x_1 - x_2) + \delta(x_1, x_2)| \\
&= |\delta(x_1, x_2)|,
\end{aligned}$$

where

$$|\delta(x_1, x_2)| \leq |x_1 - x_2| \sup |f'(z) - f'(0)| \leq s|x_1 - x_2|.$$

Hence g_y has a unique fixed point $x \in \bar{B}_r(0)$ which is such that $f(x) = y$. This proves the lemma.

We shall now prove some useful corollaries, which will be used in dealing with immersions and submersions later. *We assume that morphism means C^p-morphism with $p \geq 1$.*

Corollary 5.5. *Let U be an open subset of $\mathbf{E}$, and $f: U \to \mathbf{F}_1 \times \mathbf{F}_2$ a morphism of U into a product of vector spaces. Let $x_0 \in U$, suppose that $f(x_0) = (0, 0)$ and that $f'(x_0)$ induces a linear isomorphism of $\mathbf{E}$ and $\mathbf{F}_1 = \mathbf{F}_1 \times 0$. Then there exists a local isomorphism g of $\mathbf{F}_1 \times \mathbf{F}_2$ at $(0, 0)$ such that*

$$g \circ f: U \to \mathbf{F}_1 \times \mathbf{F}_2$$

maps an open subset U_1 of U into $\mathbf{F}_1 \times 0$ and induces a local isomorphism of U_1 at x_0 on an open neighborhood of 0 in $\mathbf{F}_1$.

Proof. We may assume without loss of generality that $\mathbf{F}_1 = \mathbf{E}$ (identify by means of $f'(x_0)$) and $x_0 = 0$. We define

$$\varphi: U \times \mathbf{F}_2 \to \mathbf{F}_1 \times \mathbf{F}_2$$

by the formula

$$\varphi(x, y_2) = f(x) + (0, y_2)$$

for $x \in U$ and $y_2 \in \mathbf{F}_2$. Then $\varphi(x, 0) = f(x)$, and

$$\varphi'(0, 0) = f'(0) + (0, \mathrm{id}_2).$$

Since $f'(0)$ is assumed to be a linear isomorphism onto $\mathbf{F}_1 \times 0$, it follows that $\varphi'(0, 0)$ is also a linear isomorphism. Hence by the theorem, it has a local inverse, say g, which obviously satisfies our requirements.

Corollary 5.6. *Let* $\mathbf{E}$, $\mathbf{F}$ *be normed vector spaces, U open in $\mathbf{E}$, and $f: U \to \mathbf{F}$ a C^p-morphism with $p \geq 1$. Let $x_0 \in U$. Suppose that $f(x_0) = 0$ and $f'(x_0)$ gives a linear isomorphism of $\mathbf{E}$ on a closed subspace of $\mathbf{F}$. Then there exists a local isomorphism $g: \mathbf{F} \to \mathbf{F}_1 \times \mathbf{F}_2$ at 0 and an open subset U_1 of U containing x_0 such that the composite map $g \circ f$ induces an isomorphism of U_1 onto an open subset of $\mathbf{F}_1$.*

Considering the splitting assumption, this is a reformulation of Corollary 5.5.

For the next corollary, dual to the preceding one, we introduce the notion of a **local projection**. Given a product of two open sets of vector spaces $V_1 \times V_2$ and a morphism $f: V_1 \times V_2 \to \mathbf{F}$, we say that f is a **projection** (on the first factor) if f can be factored

$$V_1 \times V_2 \to V_1 \to \mathbf{F}$$

into an ordinary projection and an isomorphism of V_1 onto an open subset of $\mathbf{F}$. We say that f is a local projection at (a_1, a_2) if there exists an open neighborhood $U_1 \times U_2$ of (a_1, a_2) such that the restriction of f to this neighborhood is a projection.

Corollary 5.7. *Let U be an open subset of a product of vector spaces $\mathbf{E}_1 \times \mathbf{E}_2$ and (a_1, a_2) a point of U. Let $f: U \to \mathbf{F}$ be a morphism into a Banach space, say $f(a_1, a_2) = 0$, and assume that the partial derivative*

$$D_2 f(a_1, a_2): \mathbf{E}_2 \to \mathbf{F}$$

is a linear isomorphism. Then there exists a local isomorphism h of a product $V_1 \times V_2$ onto an open neighborhood of (a_1, a_2) contained in U such that the composite map

$$V_1 \times V_2 \xrightarrow{h} U \xrightarrow{f} \mathbf{F}$$

is a projection (on the second factor).

Proof. We may assume $(a_1, a_2) = (0, 0)$ and $\mathbf{E}_2 = \mathbf{F}$. We define

$$\varphi: \mathbf{E}_1 \times \mathbf{E}_2 \to \mathbf{E}_1 \times \mathbf{E}_2$$

by

$$\varphi(x_1, x_2) = \big(x_1, f(x_1, x_2)\big)$$

locally at (a_1, a_2). Then φ' is represented by the matrix

$$\begin{pmatrix} \mathrm{id}_1 & O \\ D_1 f & D_2 f \end{pmatrix}$$

and is therefore a linear isomorphism at (a_1, a_2). By the theorem, it has a local inverse h which clearly satisfies our requirements.

Corollary 5.8. *Let U be an open subset of a vector space $\mathbf{E}$ and $f: U \to \mathbf{F}$ a morphism into a vector space $\mathbf{F}$. Let $x_0 \in U$ and assume that $f'(x_0)$ is surjective. Then there exists an open subset U' of U containing x_0 and an isomorphism*

$$h: V_1 \times V_2 \to U'$$

such that the composite map $f \circ h$ is a projection

$$V_1 \times V_2 \to V_1 \to \mathbf{F}.$$

Proof. Again this is essentially a reformulation of the corollary, taking into account the splitting assumption.

Theorem 5.9 (The Implicit Mapping Theorem). *Let U, V be open sets in normed vector spaces $\mathbf{E}$, $\mathbf{F}$ respectively, and let*

$$f: U \times V \to G$$

be a C^p mapping. Let $(a, b) \in U \times V$, and assume that

$$D_2 f(a, b): \mathbf{F} \to G$$

is a linear isomorphism. Let $f(a, b) = 0$. Then there exists a continuous map $g: U_0 \to V$ defined on an open neighborhood U_0 of a such that $g(a) = b$ and such that

$$f\big(x, g(x)\big) = 0$$

for all $x \in U_0$. If U_0 is taken to be a sufficiently small ball, then g is uniquely determined, and is also of class C^p.

Proof. Let $\lambda = D_2 f(a, b)$. Replacing f by $\lambda^{-1} \circ f$ we may assume without loss of generality that $D_2 f(a, b)$ is the identity. Consider the map

$$\varphi: \ U \times V \to E \times F$$

given by

$$\varphi(x, y) = (x, f(x, y)).$$

Then the derivative of φ at (a, b) is immediately computed to be represented by the matrix

$$D\varphi(a, b) = \begin{pmatrix} \mathrm{id}_E & O \\ D_1 f(a, b) & D_2 f(a, b) \end{pmatrix} = \begin{pmatrix} \mathrm{id}_E & O \\ D_1 f(a, b) & \mathrm{id}_F \end{pmatrix}$$

whence φ is locally invertible at (a, b) since the inverse of $D\varphi(a, b)$ exists and is the matrix

$$\begin{pmatrix} \mathrm{id}_E & O \\ -D_1 f(a, b) & \mathrm{id}_F \end{pmatrix}.$$

We denote the local inverse of φ by ψ. We can write

$$\psi(x, z) = (x, h(x, z))$$

where h is some mapping of class C^p. We define

$$g(x) = h(x, 0).$$

Then certainly g is of class C^p and

$$(x, f(x, g(x))) = \varphi(x, g(x)) = \varphi(x, h(x, 0)) = \varphi(\psi(x, 0)) = (x, 0).$$

This proves the existence of a C^p map g satisfying our requirements.

Now for the uniqueness, suppose that g_0 is a continuous map defined near a such that $g_0(a) = b$ and $f(x, g_0(x)) = c$ for all x near a. Then $g_0(x)$ is near b for such x, and hence

$$\varphi(x, g_0(x)) = (x, 0).$$

Since φ is invertible near (a, b) it follows that there is a unique point (x, y) near (a, b) such that $\varphi(x, y) = (x, 0)$. Let U_0 be a small ball on which g is defined. If g_0 is also defined on U_0, then the above argument shows that g and g_0 coincide on some smaller neighborhood of a. Let $x \in U_0$ and let $v = x - a$. Consider the set of those numbers t with $0 \leq t \leq 1$ such that $g(a + tv) = g_0(a + tv)$. This set is not empty. Let s

be its least upper bound. By continuity, we have $g(a + sv) = g_0(a + sv)$. If $s < 1$, we can apply the existence and that part of the uniqueness just proved to show that g and g_0 are in fact equal in a neighborhood of $a + sv$. Hence $s = 1$, and our uniqueness statement is proved, as well as the theorem.

Note. The particular value $f(a, b) = 0$ in the preceding theorem is irrelevant. If $f(a, b) = c$ for some $c \neq 0$, then the above proof goes through replacing 0 by c everywhere.

CHAPTER II

Manifolds

"Vector spaces" are assumed to be finite dimensional as before. Starting with open subsets of vector spaces, one can glue them together with C^p-isomorphisms. The result is called a manifold. We begin by giving the formal definition. We then make manifolds into a category, and discuss special types of morphisms. We define the tangent space at each point, and apply the criteria following the inverse function theorem to get a local splitting of a manifold when the tangent space splits at a point.

We shall wait until the next chapter to give a manifold structure to the union of all the tangent spaces.

II, §1. ATLASES, CHARTS, MORPHISMS

Let X be a Hausdorff topological space. An **atlas of class** C^p $(p \geq 0)$ on X is a collection of pairs (U_i, φ_i) (i ranging in some indexing set), satisfying the following conditions:

AT 1. *Each U_i is an open subset of X and the U_i cover X.*

AT 2. *Each φ_i is a topological isomorphism of U_i onto an open subset $\varphi_i U_i$ of some vector space $\mathbf{E}_i$ and for any i, j, $\varphi_i(U_i \cap U_j)$ is open in $\mathbf{E}_i$.*

AT 3. *The map*

$$\varphi_j \varphi_i^{-1} \colon \varphi_i(U_i \cap U_j) \to \varphi_j(U_i \cap U_j)$$

is a C^p-isomorphism for each pair of indices i, j.

Each pair (U_i, φ_i) will be called a **chart** of the atlas. If a point x of X lies in U_i, then we say that (U_i, φ_i) is a **chart at** x.

In condition **AT 2**, we did not require that the vector spaces be the same for all indices i, or even that they be linearly isomorphic. If they are all equal to the same space **E**, then we say that the atlas is an **E**-atlas. If two charts (U_i, φ_i) and (U_j, φ_j) are such that U_i and U_j have a non-empty intersection, and if $p \geq 1$, then taking the derivative of $\varphi_j \varphi_i^{-1}$ we see that $\mathbf{E}_i$ and $\mathbf{E}_j$ are linearly isomorphic. Furthermore, the set of points $x \in X$ for which there exists a chart (U_i, φ_i) at x such that $\mathbf{E}_i$ is linearly isomorphic to a given space **E** is both open and closed. Consequently, on each connected component of X, we could assume that we have an **E**-atlas for some fixed **E**.

Suppose that we are given an open subset U of X and a topological isomorphism $\varphi: U \to U'$ onto an open subset of some vector space **E**. We shall say that (U, φ) is **compatible** with the atlas $\{(U_i, \varphi_i)\}$ if each map $\varphi_i \varphi^{-1}$ (defined on a suitable intersection as in **AT 3**) is a C^p-isomorphism. Two atlases are said to be **compatible** if each chart of one is compatible with the other atlas. One verifies immediately that the relation of compatibility between atlases is an equivalence relation. An equivalence class of atlases of class C^p on X is said to define a structure of C^p-**manifold** on X. If all the vector spaces $\mathbf{E}_i$ in some atlas are linearly isomorphic, then we can always find an equivalent atlas for which they are all equal, say to the vector space **E**. We then say that X is an **E-manifold** or that X is **modeled** on **E**.

If $\mathbf{E} = \mathbf{R}^n$ for some fixed n, then we say that the manifold is n-**dimensional**. In this case, a chart

$$\varphi: \ U \to \mathbf{R}^n$$

is given by n coordinate functions $\varphi_1, \ldots, \varphi_n$. If P denotes a point of U, these functions are often written

$$x_1(P), \ldots, x_n(P),$$

or simply $x_1, \ldots, x_n$. They are called **local coordinates** on the manifold.

If the integer p (which may also be ∞) is fixed throughout a discussion, we also say that X is a manifold.

The collection of C^p-manifolds will be denoted by Man^p. We shall make these into categories by defining morphisms below.

Let X be a manifold, and U an open subset of X. Then it is possible, in the obvious way, to induce a manifold structure on U, by taking as charts the intersections

$$\left(U_i \cap U, \ \varphi_i | (U_i \cap U) \right).$$

If X is a topological space, covered by open subsets V_j, and if we are given on each V_j a manifold structure such that for each pair j, j' the induced structure on $V_j \cap V_{j'}$ coincides, then it is clear that we can give to X a unique manifold structure inducing the given ones on each V_j.

Example. Let X be the real line, and for each open interval U_i, let φ_i be the function $\varphi_i(t) = t^3$. Then the $\varphi_j \varphi_i^{-1}$ are all equal to the identity, and thus we have defined a C^∞-manifold structure on **R**!

If X, Y are two manifolds, then one can give the product $X \times Y$ a manifold structure in the obvious way. If $\{(U_i, \varphi_i)\}$ and $\{(V_j, \psi_j)\}$ are atlases for X, Y respectively, then

$$\{(U_i \times V_j, \varphi_i \times \psi_j)\}$$

is an atlas for the product, and the product of compatible atlases gives rise to compatible atlases, so that we do get a well-defined product structure.

Let X, Y be two manifolds. Let $f \colon X \to Y$ be a map. We shall say that f is a C^p-**morphism** if, given $x \in X$, there exists a chart (U, φ) at x and a chart (V, ψ) at $f(x)$ such that $f(U) \subset V$, and the map

$$\psi \circ f \circ \varphi^{-1} \colon \varphi U \to \psi V$$

is a C^p-morphism in the sense of Chapter I, §3. One sees then immediately that this same condition holds for any choice of charts (U, φ) at x and (V, ψ) at $f(x)$ such that $f(U) \subset V$.

It is clear that the composite of two C^p-morphisms is itself a C^p-morphism (because it is true for open subsets of vector spaces). The C^p-manifolds and C^p-morphisms form a category. The notion of isomorphism is therefore defined, and we observe that in our example of the real line, the map $t \mapsto t^3$ gives an isomorphism between the funny differentiable structure and the usual one.

If $f \colon X \to Y$ is a morphism, and (U, φ) is a chart at a point $x \in X$, while (V, ψ) is a chart at $f(x)$, then we shall also denote by

$$f_{V, U} \colon \varphi U \to \psi V$$

the map $\psi f \varphi^{-1}$.

It is also convenient to have a local terminology. Let U be an open set (of a manifold or a Banach space) containing a point x_0. By a **local isomorphism** at x_0 we mean an isomorphism

$$f \colon U_1 \to V$$

from some open set U_1 containing x_0 (and contained in U) to an open set V (in some manifold or some vector space). Thus a local isomorphism is essentially a change of chart, locally near a given point.

II, §2. SUBMANIFOLDS, IMMERSIONS, SUBMERSIONS

Let X be a topological space, and Y a subset of X. We say that Y is **locally closed** in X if every point $y \in Y$ has an open neighborhood U in X such that $Y \cap U$ is closed in U. One verifies easily that a locally closed subset is the intersection of an open set and a closed set. For instance, any open subset of X is locally closed, and any open interval is locally closed in the plane.

Let X be a manifold (of class C^p with $p \geq 0$). Let Y be a subset of X and assume that for each point $y \in Y$ there exists a chart (V, ψ) at y such that ψ gives an isomorphism of V with a product $V_1 \times V_2$ where V_1 is open in some space $\mathbf{E}_1$ and V_2 is open in some space $\mathbf{E}_2$, and such that

$$\psi(Y \cap V) = V_1 \times a_2$$

for some point $a_2 \in V_2$ (which we could take to be 0). Then it is clear that Y is locally closed in X. Furthermore, the map ψ induces a bijection

$$\psi_1 : Y \cap V \to V_1.$$

The collection of pairs $(Y \cap V, \psi_1)$ obtained in the above manner constitutes an atlas for Y, of class C^p. The verification of this assertion, whose formal details we leave to the reader, depends on the following obvious fact.

Lemma 2.1. *Let U_1, U_2, V_1, V_2 be open subsets of vector spaces, and $g \colon U_1 \times U_2 \to V_1 \times V_2$ a C^p-morphism. Let $a_2 \in U_2$ and $b_2 \in V_2$ and assume that g maps $U_1 \times a_2$ into $V_1 \times b_2$. Then the induced map*

$$g_1 \colon U_1 \to V_1$$

is also a morphism.

Indeed, it is obtained as a composite map

$$U_1 \to U_1 \times U_2 \to V_1 \times V_2 \to V_1,$$

the first map being an inclusion and the third a projection.

We have therefore defined a C^p-structure on Y which will be called a **submanifold** of X. This structure satisfies a universal mapping property, which characterizes it, namely:

Given any map $f \colon Z \to X$ from a manifold Z into X such that $f(Z)$ is contained in Y. Let $f_Y \colon Z \to Y$ be the induced map. Then f is a morphism if and only if f_Y is a morphism.

The proof of this assertion depends on Lemma 2.1, and is trivial.

Finally, we note that the inclusion of Y into X is a morphism.
If Y is also a closed subspace of X, then we say that it is a **closed submanifold**.

Suppose that X is a manifold of dimension n, and that Y is a submanifold of dimension r. Then from the definition we see that the local product structure in a neighborhood of a point of Y can be expressed in terms of local coordinates as follows. Each point P of Y has an open neighborhood U in X with local coordinates $(x_1, \ldots, x_n)$ such that the points of Y in U are precisely those whose last $n - r$ coordinates are 0, that is, those points having coordinates of type

$$(x_1, \ldots, x_r, 0, \ldots, 0).$$

Let $f: Z \to X$ be a morphism, and let $z \in Z$. We shall say that f is an **immersion** at z if there exists an open neighborhood Z_1 of z in Z such that the restriction of f to Z_1 induces an isomorphism of Z_1 onto a submanifold of X. We say that f is an **immersion** if it is an immersion at every point.

Note that there exist injective immersions which are not isomorphisms onto submanifolds, as given by the following example:

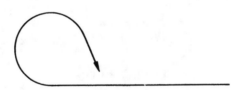

(The arrow means that the line approaches itself without touching.) An immersion which does give an isomorphism onto a submanifold is called an **embedding**, and it is called a **closed embedding** if this submanifold is closed.

A morphism $f: X \to Y$ will be called a **submersion** at a point $x \in X$ if there exists a chart (U, φ) at x and a chart (V, ψ) at $f(x)$ such that φ gives an isomorphism of U on a products $U_1 \times U_2$ (U_1 and U_2 open in some vector spaces), and such that the map

$$\psi f \varphi^{-1} = f_{V,U}: \ U_1 \times U_2 \to V$$

is a projection. One sees then that the image of a submersion is an open subset (a submersion is in fact an open mapping). We say that f is a **submersion** if it is a submersion at every point.

We have the usual criterion for immersions and submersions in terms of the derivative.

Proposition 2.2. *Let X, Y be manifolds of class C^p $(p \geq 1)$. Let $f: X \to Y$ be a C^p-morphism. Let $x \in X$. Then:*

(i) *f is an immersion at x if and only if there exists a chart (U, φ) at x and (V, ψ) at $f(x)$ such that $f'_{V,U}(\varphi x)$ is injective.*
(ii) *f is a submersion at x if and only if there exists a chart (U, φ) at x and (V, ψ) at $f(x)$ such that $f'_{V,U}(\varphi x)$ is surjective.*

Proof. This is an immediate consequence of Corollaries 5.4 and 5.6 of the inverse mapping theorem.

The conditions expressed in (i) and (ii) depend only on the derivative, and if they hold for one choice of charts (U, φ) and (V, ψ) respectively, then they hold for every choice of such charts. It is therefore convenient to introduce a terminology in order to deal with such properties.

Let X be a manifold of class C^p $(p \geq 1)$. Let x be a point of X. We consider triples (U, φ, v) where (U, φ) is a chart at x and v is an element of the vector space in which φU lies. We say that two such triples (U, φ, v) and (V, ψ, w) are **equivalent** if the derivative of $\psi \varphi^{-1}$ at φx maps v on w. The formula reads:

$$(\psi \varphi^{-1})'(\varphi x)v = w$$

(obviously an equivalence relation by the chain rule). An equivalence class of such triples is called a **tangent vector** of X at x. The set of such tangent vectors is called the **tangent space** of X at x and is denoted by $T_x(X)$. Each chart (U, φ) determines a bijection of $T_x(X)$ on a vector space, namely the equivalence class of (U, φ, v) corresponds to the vector v. By means of such a bijection it is possible to transport to $T_x(X)$ the structure of vector space given by the chart, and it is immediate that this structure is independent of the chart selected.

If U, V are open in vector spaces, then to every morphism of class C^p $(p \geq 1)$ we can associate its derivative $Df(x)$. If now $f: X \to Y$ is a morphism of one manifold into another, and x a point of X, then by means of charts we can interpret the derivative of f on each chart at x as a mapping

$$df(x) = T_x f: T_x(X) \to T_{f(x)}(Y).$$

Indeed, this map $T_x f$ is the unique linear map having the following property. If (U, φ) is a chart at x and (V, ψ) is a chart at $f(x)$ such that $f(U) \subset V$ and $\bar{v}$ is a tangent vector at x represented by v in the chart

(U, φ), then

$$T_x f(\bar{v})$$

is the tangent vector at $f(x)$ represented by $Df_{V, U}(x)v$. The representation of $T_x f$ on the spaces of charts can be given in the form of a diagram

$$
\begin{array}{ccc}
T_x(X) & \longrightarrow & \mathbf{E} \\
{\scriptstyle T_x f}\big\downarrow & & \big\downarrow {\scriptstyle f'_{V, U}(x)} \\
T_{f(x)}(Y) & \longrightarrow & \mathbf{F}
\end{array}
$$

The map $T_x f$ is obviously linear.

As a matter of notation, we shall sometimes write $f_{*, x}$ instead of $T_x f$.

The operation T satisfies an obvious functorial property, namely, if $f: X \to Y$ and $g: Y \to Z$ are morphisms, then

$$T_x(g \circ f) = T_{f(x)}(g) \circ T_x(f),$$

$$T_x(\mathrm{id}) = \mathrm{id}.$$

We may reformulate Proposition 2.2:

Proposition 2.3. *Let X, Y be manifolds of class C^p $(p \geq 1)$. Let $f: X \to Y$ be a C^p-morphism. Let $x \in X$. Then:*

(i) *f is an immersion at x if and only if the map $T_x f$ is injective.*
(ii) *f is a submersion at x if and only if the map $T_x f$ is surjective.*

Example. Let $\mathbf{E}$ be a vector space with positive definite scalar product, and let $\langle x, y \rangle \in \mathbf{R}$ be its scalar product. Then the square of the norm $f(x) = \langle x, x \rangle$ is obviously of class C^∞. The derivative $f'(x)$ is given by the formula

$$f'(x)y = 2\langle x, y \rangle$$

and for any given $x \neq 0$, it follows that the derivative $f'(x)$ is surjective. Furthermore, its kernel is the orthogonal complement of the subspace generated by x. Consequently the unit sphere in euclidean space is a submanifold.

If W is a submanifold of a manifold Y of class C^p $(p \geq 1)$, then the inclusion

$$i: W \to Y$$

induces a map

$$T_w i: \ T_w(W) \to T_w(Y)$$

which is in fact an injection. It will be convenient to identify $T_w(W)$ in $T_w(Y)$ if no confusion can result.

A morphism $f: X \to Y$ will be said to be **transversal** over the submanifold W of Y if the following condition is satisfied.

Let $x \in X$ be such that $f(x) \in W$. Let (V, ψ) be a chart at $f(x)$ such that $\psi: V \to V_1 \times V_2$ is an isomorphism on a product, with

$$\psi\big(f(x)\big) = (0, 0) \qquad \text{and} \qquad \psi(W \cap V) = V_1 \times 0.$$

Then there exists an open neighborhood U of x such that the composite map

$$U \xrightarrow{f} V \xrightarrow{\psi} V_1 \times V_2 \xrightarrow{\text{pr}} V_2$$

is a submersion.

In particular, if f is transversal over W, then $f^{-1}(W)$ is a submanifold of X, because the inverse image of 0 by our local composite map

$$\text{pr} \circ \psi \circ f$$

is equal to the inverse image of $W \cap V$ by ψ.

As with immersions and submersions, we have a characterization of transversal maps in terms of tangent spaces.

Proposition 2.4. *Let* X, Y *be manifolds of class* C^p $(p \geq 1)$. *Let* $f: X \to Y$ *be a* C^p-*morphism, and* W *a submanifold of* Y. *The map* f *is transversal over* W *if and only if for each* $x \in X$ *such that* $f(x)$ *lies in* W, *the composite map*

$$T_x(X) \xrightarrow{T_x f} T_w(Y) \to T_w(Y)/T_w(W)$$

with $w = f(x)$ *is surjective.*

Proof. If f is transversal over W, then for each point $x \in X$ such that $f(x)$ lies in W, we choose charts as in the definition, and reduce the question to one of maps of open subsets of vector spaces. In that case, the conclusion concerning the tangent spaces follows at once from the assumed direct product decompositions. Conversely, assume our condition on the tangent map. The question being local, we can assume that $Y = V_1 \times V_2$ is a product of open sets in vector spaces such that $W = V_1 \times 0$, and we can also assume that $X = U$ is open in some vector space, $x = 0$. Then we let $g: U \to V_2$ be the map $\pi \circ f$ where π is the projection, and note that our assumption means that $g'(0)$ is surjective. Furthermore, $g^{-1}(0) = f^{-1}(W)$. We can then use Corollary 5.7 of the inverse mapping theorem to conclude the proof.

Remark. In the statement of our proposition, we observe that the surjectivity of the composite map is equivalent to the fact that $T_w(Y)$ is equal to the sum of the image of $T_x f$ and $T_w(W)$, that is

$$T_w(Y) = \operatorname{Im}(T_x f) + \operatorname{Im}(T_x i),$$

where $i\colon W \to Y$ is the inclusion.

If $\mathbf{E}$ is a vector space, then the diagonal Δ in $\mathbf{E} \times \mathbf{E}$ is a closed subspace. Either factor $\mathbf{E} \times 0$ or $0 \times \mathbf{E}$ is a closed complement. Consequently, the diagonal is a closed submanifold of $\mathbf{E} \times \mathbf{E}$. If X is any manifold of class C^p, $p \geq 1$, then the diagonal is therefore also a submanifold.

Let $f\colon X \to Z$ and $g\colon Y \to Z$ be two C^p-morphisms, $p \geq 1$. We say that they are **transversal** if the morphism

$$f \times g\colon X \times Y \to Z \times Z$$

is transversal over the diagonal. We remark right away that the surjectivity of the map in Proposition 2.4 can be expressed in two ways. Given two points $x \in X$ and $y \in Y$ such that $f(x) = g(y) = z$, the condition

$$\operatorname{Im}(T_x f) + \operatorname{Im}(T_y g) = T_z(Z)$$

is equivalent to the condition

$$\operatorname{Im}\left(T_{(x,y)}(f \times g)\right) + T_{(z,z)}(\Delta) = T_{(z,z)}(Z \times Z).$$

Thus in the finite dimensional case, we could take it as definition of transversality.

We use transversality as a sufficient condition under which the fiber product of two morphisms exists. We recall that in any category, the **fiber product** of two morphisms $f\colon X \to Z$ and $g\colon Y \to Z$ over Z consists of an object P and two morphisms

$$g_1\colon P \to X \qquad \text{and} \qquad g_2\colon P \to Y$$

such that $f \circ g_1 = g \circ g_2$, and satisfying the universal mapping property: Given an object S and two morphisms $u_1\colon S \to X$ and $u_2\colon S \to Y$ such that $f u_1 = g u_2$, there exists a unique morphism $u\colon S \to P$ making the following diagram commutative:

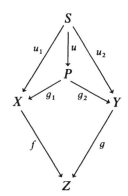

The triple (P, g_1, g_2) is uniquely determined, up to a unique isomorphism (in the obvious sense), and P is also denoted by $X \times_Z Y$.

One can view the fiber product unsymmetrically. Given two morphisms f, g as in the following diagram:

$$
\begin{array}{ccc}
 & & Y \\
 & & \downarrow g \\
X & \xrightarrow{\;\;f\;\;} & Z
\end{array}
$$

assume that their fiber product exists, so that we can fill in the diagram:

$$
\begin{array}{ccc}
X \times_Z Y & \longrightarrow & Y \\
\downarrow g_1 & & \downarrow g \\
X & \longrightarrow & Z
\end{array}
$$

We say that g_1 is the **pull back** of g by f, and also write it as $f^*(g)$. Similarly, we write $X \times_Z Y$ as $f^*(Y)$.

In our category of manifolds, we shall deal only with cases when the fiber product can be taken to be the set-theoretic fiber product on which a manifold structure has been defined. (The set-theoretic fiber product is the set of pairs of points projecting on the same point.) This determines the fiber product uniquely, and not only up to a unique isomorphism.

Proposition 2.5. *Let $f: X \to Z$ and $g: Y \to Z$ be two C^p-morphisms with $p \geqq 1$. If they are transversal, then*

$$
(f \times g)^{-1}(\Delta_Z),
$$

together with the natural morphisms into X and Y (obtained from the projections), is a fiber product of f and g over Z.

Proof. Obvious.

To construct a fiber product, it suffices to do it locally. Indeed, let $f: X \to Z$ and $g: Y \to Z$ be two morphisms. Let $\{V_i\}$ be an open covering of Z, and let

$$f_i: f^{-1}(V_i) \to V_i \quad \text{and} \quad g_i: g^{-1}(V_i) \to V_i$$

be the restrictions of f and g to the respective inverse images of V_i. Let $P = (f \times g)^{-1}(\Delta_Z)$. Then P consists of the points (x, y) with $x \in X$ and $y \in Y$ such that $f(x) = g(y)$. We view P as a subspace of $X \times Y$ (i.e. with the topology induced by that of $X \times Y$). Similarly, we construct P_i with f_i and g_i. Then P_i is open in P. The projections on the first and second factors give natural maps of P_i into $f^{-1}(V_i)$ and $g^{-1}(V_i)$ and of P into X and Y.

Proposition 2.6. *Assume that each P_i admits a manifold structure (compatible with its topology) such that these maps are morphisms, making P_i into a fiber product of f_i and g_i. Then P, with its natural projections, is a fiber product of f and g.*

To prove the above assertion, we observe that the P_i form a covering of P. Furthermore, the manifold structure on $P_i \cap P_j$ induced by that of P_i or P_j must be the same, because it is the unique fiber product structure over $V_i \cap V_j$, for the maps f_{ij} and g_{ij} (defined on $f^{-1}(V_i \cap V_j)$ and $g^{-1}(V_i \cap V_j)$ respectively). Thus we can give P a manifold structure, in such a way that the two projections into X and Y are morphisms, and make P into a fiber product of f and g.

We shall apply the preceding discussion to vector bundles in the next chapter, and the following local criterion will be useful.

Proposition 2.7. *Let $f: X \to Z$ be a morphism, and $g: Z \times W \to Z$ be the projection on the first factor. Then f, g have a fiber product, namely the product $X \times W$ together with the morphisms of the following diagram:*

$$
\begin{array}{ccc}
X \times W & \xrightarrow{f \times \mathrm{id}} & Z \times W \\
{\scriptstyle \mathrm{pr}_1} \downarrow & & \downarrow {\scriptstyle \mathrm{pr}_1} \\
X & \xrightarrow{\quad f \quad} & Z
\end{array}
$$

II, §3. PARTITIONS OF UNITY

Let X be a manifold of class C^p. A **function** on X will be a morphism of X into $\mathbf{R}$, of class C^p, unless otherwise specified. The C^p functions form a ring denoted by $\mathfrak{F}^p(X)$ or $\mathrm{Fu}^p(X)$. The **support** of a function f is the closure of the set of points x such that $f(x) \neq 0$.

Let X be a topological space. A covering of X is **locally finite** if every point has a neighborhood which intersects only finitely many elements of the covering. A **refinement** of a covering of X is a second covering, each element of which is contained in an element of the first covering. A topological space is **paracompact** if it is Hausdorff, and every open covering has a locally finite open refinement.

Proposition 3.1. *If X is a paracompact space, and if $\{U_i\}$ is an open covering, then there exists a locally finite open covering $\{V_i\}$ such that $V_i \subset U_i$ for each i.*

Proof. Let $\{V_k\}$ be a locally finite open refinement of $\{U_i\}$. For each k there is an index $i(k)$ such that $V_k \subset U_{i(k)}$. We let W_i be the union of those V_k such that $i(k) = i$. Then the W_i form a locally finite open covering, because any neighborhood of a point which meets infinitely many W_i must also meet infinitely many V_k.

Proposition 3.2. *If X is paracompact, then X is normal. If, furthermore, $\{U_i\}$ is a locally finite open covering of X, then there exists a locally finite open covering $\{V_i\}$ such that $\overline{V}_i \subset U_i$.*

Proof. We refer the reader to Bourbaki [Bou 68].

Observe that Proposition 3.1 shows that the insistence that the indexing set of a refinement be a given one can easily be achieved.

A **partition of unity** (of class C^p) on a manifold X consists of an open covering $\{U_i\}$ of X and a family of functions

$$\psi_i \colon X \to \mathbf{R}$$

satisfying the following conditions:

PU 1. *For all $x \in X$ we have $\psi_i(x) \geq 0$.*

PU 2. *The support of ψ_i is contained in U_i.*

PU 3. *The covering is locally finite.*

PU 4. *For each point $x \in X$ we have*

$$\sum \psi_i(x) = 1.$$

(The sum is taken over all i, but is in fact finite for any given point x in view of **PU 3**.)

We sometimes say that $\{(U_i, \psi_i)\}$ is a partition of unity.

A manifold X will be said to **admit partitions of unity** if it is paracompact, and if, given a locally finite open covering $\{U_i\}$, there exists a partition of unity $\{\psi_i\}$ such that the support of ψ_i is contained in U_i.

If $\{U_i\}$ is a covering of X, then we say that a covering $\{V_k\}$ is subordinated to $\{U_i\}$ if each V_k is contained in some U_i.

It is desirable to give sufficient conditions on a manifold in order to insure the existence of partitions of unity. There is no difficulty with the topological aspects of this problem. It is known that a metric space is paracompact (cf. Bourbaki [Bou 68], [Ke 55]), and on a paracompact space, one knows how to construct continuous partitions of unity (loc. cit.).

If E is a euclidean space, we denote by $B_r(a)$ the open ball of radius r and center a, and by $\bar{B}_r(a)$ the closed ball of radius r and center a. If $a = 0$, then we write B_r and $\bar{B}_r$ respectively. Two open balls (of finite radius) are obviously C^∞-isomorphic. If X is a manifold and (V, φ) is a chart at a point $x \in X$, then we say that (V, φ) (or simply V) is a ball of radius r if φV is a ball of radius r. We now use euclidean space for charts, with the given euclidean norm.

Theorem 3.3. *Let X be a manifold whose topology has a countable base. Given an open covering of X, then there exists an atlas $\{(V_k, \varphi_k)\}$ such that the covering $\{V_k\}$ is locally finite and subordinated to the given covering, such that $\varphi_k V_k$ is the open ball B_3, and such that the open sets $W_k = \varphi_k^{-1}(B_1)$ cover X.*

Proof. Let $U_1, U_2, \ldots$ be a basis for the open sets of X such that each $\bar{U}_i$ is compact. We construct inductively a sequence $A_1, A_2, \ldots$ of compact sets whose union is X, such that A_i is contained in the interior of A_{i+1}. We let $A_1 = \bar{U}_1$. Suppose we have constructed A_i. We let j be the smallest integer such that A_i is contained in $U_1 \cup \cdots \cup U_j$. We let A_{i+1} be the closed and compact set

$$\bar{U}_1 \cup \cdots \cup \bar{U}_j \cup \bar{U}_{i+1}.$$

For each point $x \in X$ we can find an arbitrarily small chart (V_x, φ_x) at x such that $\varphi_x V_x$ is the ball of radius 3 (so that each V_x is contained in some element of U). We let $W_x = \varphi_x^{-1}(B_1)$ be the ball of radius 1 in this chart. We can cover the set

$$A_{i+1} - \text{Int}(A_i)$$

(intuitively the closed annulus) by a finite number of these balls of radius 1, say $W_1, \ldots, W_n$, such that, at the same time, each one of $V_1, \ldots, V_n$ is contained in the open set $\text{Int}(A_{i+2}) - A_{i-1}$ (intuitively, the open annulus of the next bigger size). We let $\mathfrak{B}_i$ denote the collection $V_1, \ldots, V_n$ and let $\mathfrak{B}$ be composed of the union of the $\mathfrak{B}_i$. Then $\mathfrak{B}$ is locally finite, and we are done.

Corollary 3.4. *Let X be a manifold whose topology has a countable base. Then X admits partitions of unity.*

Proof. Let $\{(V_k, \varphi_k)\}$ be as in the theorem, and $W_k = \varphi_k^{-1}(B_1)$. We can find a function ψ_k of class C^p such that $0 \leq \psi_k \leq 1$, such that $\psi_k(x) = 1$ for $x \in W_k$ and $\psi_k(x) = 0$ for $x \notin V_k$. (The proof is recalled below.) We now let

$$\psi = \sum \psi_k$$

(a sum which is finite at each point), and we let $\gamma_k = \psi_k/\psi$. Then $\{(V_k, \gamma_k)\}$ is the desired partition of unity.

We now recall the argument giving the function ψ_k. First, given two real numbers r, s with $0 \leq r < s$, the function defined by

$$\exp\left(\frac{-1}{(t-r)(s-t)}\right)$$

in the open interval $r < t < s$ and 0 outside the interval determines a bell-shaped C^∞-function from $\mathbf{R}$ into $\mathbf{R}$. Its integral from minus infinity to t, divided by the area under the bell yields a function which lies strictly between 0 and 1 in the interval $r < t < s$, is equal to 0 for $t \leq r$ and is equal to 1 for $t \geq s$. (The function is even monotone increasing.)

We can therefore find a real valued function of a real variable, say $\eta(t)$, such that $\eta(t) = 1$ for $|t| < 1$ and $\eta(t) = 0$ for $|t| \geq 1 + \delta$ with small δ, and such that $0 \leq \eta \leq 1$. If $\mathbf{E}$ is a euclidean space, then $\eta(|x|^2) = \psi(x)$ gives us a function which is equal to 1 on the ball of radius 1 and 0 outside the ball of radius $1 + \delta$. This function can then be transported to the manifold by any given chart whose image is the ball of radius 3. For convenience, we state separately what we have just proved.

Lemma 3.5. *Let E be a euclidean space. There exists a C^∞ real function ψ on E such that $\psi(x) = 1$ for $|x| \leq 1$, $\psi(x) > 0$ for $|x| < 1 + \delta$, and $\psi(x) = 0$ for $|x| \geq 1 + \delta$. Alternatively, there exists a C^∞ function h such that*

$$h(x) > 0 \quad \text{for} \quad |x| < 1 \quad \text{and} \quad h(x) = 0 \quad \text{for} \quad |x| \geq 1.$$

In other words, one would construct a function which is > 0 on a given ball and $= 0$ outside this ball.

Partitions of unity constitute the only known means of gluing together local mappings (into objects having an addition, namely vector bundles, discussed in the next chapter).

II, §4. MANIFOLDS WITH BOUNDARY

Let **E** be a vector space, and $\lambda\colon \mathbf{E} \to \mathbf{R}$ a linear map into **R**. (This will also be called a **functional** on **E**.) We denote by $\mathbf{E}_\lambda^0$ the kernel of λ, and by $\mathbf{E}_\lambda^+$ (resp. $\mathbf{E}_\lambda^-$) the set of points $x \in \mathbf{E}$ such that $\lambda(x) \geqq 0$ (resp. $\lambda(x) \leqq 0$). We call $\mathbf{E}_\lambda^0$ a **hyperplane** and $\mathbf{E}_\lambda^+$ or $\mathbf{E}_\lambda^-$ a **half plane**.

If μ is another functional and $\mathbf{E}_\lambda^+ = \mathbf{E}_\mu^+$, then there exists a number $c > 0$ such that $\lambda = c\mu$. This is easily proved. Indeed, we see at once that the kernels of λ and μ must be equal. Suppose $\lambda \neq 0$. Let x_0 be such that $\lambda(x_0) > 0$. Then $\mu(x_0) > 0$ also. The functional

$$\lambda - \big(\lambda(x_0)/\mu(x_0)\big)\mu$$

vanishes on the kernel of λ (or μ) and also on x_0. Therefore it is the 0 functional, and $c = \lambda(x_0)/\mu(x_0)$.

Let **E**, **F** be vector spaces, and let $\mathbf{E}_\lambda^+$ and $\mathbf{F}_\mu^+$ be two half planes in **E** and **F** respectively. Let U, V be two open subsets of these half planes respectively. We shall say that a mapping

$$f\colon U \to V$$

is a morphism of class C^p if the following condition is satisfied. Given a point $x \in U$, there exists an open neighborhood U_1 of x in **E**, an open neighborhood V_1 of $f(x)$ in **F**, and a morphism $f_1\colon U_1 \to V_1$ (in the sense of Chapter I) such that the restriction of f_1 to $U_1 \cap U$ is equal to f. (We assume that all morphisms are of class C^p with $p \geqq 1$.)

If our half planes are full planes (i.e. equal to the vector spaces themselves), then our present definition is the same as the one used previously.

If we take as objects the open subsets of half planes in vector spaces, and as morphisms the C^p-morphisms, then we obtain a category. The notion of isomorphism is therefore defined, and the definition of manifold by means of atlases and charts can be used as before. The manifolds of §1 should have been called **manifolds without boundary**, reserving the name of manifold for our new globalized objects. However, in most of this book, we shall deal exclusively with manifolds without boundary for simplicity. The following remarks will give readers the means of extending any result they wish (provided it is true) for the case of manifolds without boundaries to the case manifolds with.

First, concerning the notion of derivative, we have:

Proposition 4.1. *Let $f: U \to F$ and $g: U \to F$ be two morphisms of class C^p $(p \geqq 1)$ defined on an open subset U of E. Assume that f and g have the same restriction to $U \cap E_\lambda^+$ for some half plane E_λ^+, and let*

$$x \in U \cap E_\lambda^+.$$

Then $f'(x) = g'(x)$.

Proof. After considering the difference of f and g, we may assume without loss of generality that the restriction of f to $U \cap E_\lambda^+$ is 0. It is then obvious that $f'(x) = 0$.

Proposition 4.2. *Let U be open in E. Let μ be a non-zero functional on F and let $f: U \to F_\mu^+$ be a morphism of class C^p with $p \geqq 1$. If x is a point of U such that $f(x)$ lies in F_μ^0 then $f'(x)$ maps E into F_μ^0.*

Proof. Without loss of generality, we may assume that $x = 0$ and $f(x) = 0$. Let W be a given neighborhood of 0 in F. Suppose that we can find a small element $v \in E$ such that $\mu f'(0)v \neq 0$. We can write (for small t):

$$f(tv) = tf'(0)v + o(t)w_t$$

with some element $w_t \in W$. By assumption, $f(tv)$ lies in F_μ^+. Applying μ we get

$$t\mu f'(0)v + o(t)\mu(w_t) \geqq 0.$$

Dividing by t, this yields

$$\mu f'(0)v \geqq \frac{o(t)}{t}\mu(w_t).$$

Replacing t by $-t$, we get a similar inequality on the other side. Letting t tend to 0 shows that $\mu f'(0)v = 0$, a contradiction.

Let U be open in some half plane E_λ^+. We define the **boundary** of U (written ∂U) to be the intersection of U with E_λ^0, and the **interior** of U (written $\mathrm{Int}(U)$) to be the complement of ∂U in U. Then $\mathrm{Int}(U)$ is open in E.

It follows at once from our definition of differentiability that a half plane is C^∞-isomorphic with a product

$$E_\lambda^+ \approx E_\lambda^0 \times \mathbf{R}^+$$

where $\mathbf{R}^+$ is the set of real numbers $\geqq 0$, whenever $\lambda \neq 0$. The boundary of $\mathbf{E}_\lambda^+$ in that case is $\mathbf{E}_\lambda^0 \times 0$.

Proposition 4.3. *Let λ be a functional on $\mathbf{E}$ and μ a functional on $\mathbf{F}$. Let U be open in $\mathbf{E}_\lambda^+$ and V open in $\mathbf{F}_\mu^+$ and assume $U \cap \mathbf{E}_\lambda^0$, $V \cap \mathbf{F}_\mu^0$ are not empty. Let $f \colon U \to V$ be an isomorphism of class C^p $(p \geq 1)$. Then $\lambda \neq 0$ if and only if $\mu \neq 0$. If $\lambda \neq 0$, then f induces a C^p-isomorphism of* $\mathrm{Int}(U)$ *on* $\mathrm{Int}(V)$ *and of ∂U on ∂V.*

Proof. By the functoriality of the derivative, we know that $f'(x)$ is a toplinear isomorphism for each $x \in U$. Our first assertion follows from the preceding proposition. We also see that no interior point of U maps on a boundary point of V and conversely. Thus f induces a bijection of ∂U on ∂V and a bijection of $\mathrm{Int}(U)$ on $\mathrm{Int}(V)$. Since these interiors are open in their respective spaces, our definition of derivative shows that f induces an isomorphism between them. As for the boundary, it is a submanifold of the full space, and locally, our definition of derivative, together with the product structure, shows that the restriction of f to ∂U must be an isomorphism on ∂V.

This last proposition shows that the boundary is a differentiable invariant, and thus that we can speak of the boundary of a manifold.

We give just two words of warning concerning manifolds with boundary. First, products do not exist in their category. Indeed, to get products, we are forced to define manifolds with **corners**, which would take us too far afield.

Second, in defining immersions or submanifolds, there is a difference in kind when we consider a manifold embedded in a manifold without boundary, or a manifold embedded in another manifold with boundary. Think of a closed interval embedded in an ordinary half plane. Two cases arise. The case where the interval lies inside the interior of the half plane is essentially distinct from the case where the interval has one end point touching the hyperplane forming the boundary of the half plane. (For instance, given two embeddings of the first type, there exists an automorphism of the half plane carrying one into the other, but there cannot exist an automorphism of the half plane carrying an embedding of the first type into one of the second type.)

We leave it to the reader to go systematically through the notions of tangent space, immersion, embedding (and later, tangent bundle, vector field, etc.) for arbitrary manifolds (with boundary). For instance, Proposition 2.2 shows at once how to get the tangent space functorially.

CHAPTER III

Vector Bundles

The collection of tangent spaces can be glued together to give a manifold with a natural projection, thus giving rise to the tangent bundle. The general glueing procedure can be used to construct more general objects known as vector bundles, which give powerful invariants of a given manifold. (For an interesting theorem see Mazur [Maz 61].) In this chapter, we develop purely formally certain functorial constructions having to do with vector bundles. In the chapters on differential forms and Riemannian metrics, we shall discuss in greater details the constructions associated with multilinear alternating forms, and symmetric positive definite forms.

Partitions of unity are an essential tool when considering vector bundles. They can be used to combine together a random collection of morphisms into vector bundles, and we shall give a few examples showing how this can be done (concerning exact sequences of bundles).

III, §1. DEFINITION, PULL BACKS

Let X be a manifold (of class C^p with $p \geq 0$) and let $\pi: E \to X$ be a morphism. Let $\mathbf{E}$ be a vector space (always assumed finite dimensional).

Let $\{U_i\}$ be an open covering of X, and for each i, suppose that we are given a mapping

$$\tau_i: \pi^{-1}(U_i) \to U_i \times \mathbf{E}$$

satisfying the following conditions:

VB 1. *The map τ_i is a C^p isomorphism commuting with the projection on U_i, that is, such that the following diagram is commutative:*

$$\pi^{-1}(U_i) \xrightarrow{\ \tau_i\ } U_i \times \mathbf{E}$$
$$U_i$$

In particular, we obtain an isomorphism on each fiber (written $\tau_i(x)$ or τ_{ix})

$$\tau_{ix}\colon\ \pi^{-1}(x) \to \{x\} \times \mathbf{E}$$

VB 2. *For each pair of open sets U_i, U_j the map*

$$\tau_{jx} \circ \tau_{ix}^{-1}\colon\ \mathbf{E} \to \mathbf{E}$$

is a toplinear isomorphism.

VB 3. *If U_i and U_j are two members of the covering, then the map of $U_i \cap U_j$ into $L(\mathbf{E}, \mathbf{E})$ (actually $\mathrm{Laut}(\mathbf{E})$) given by*

$$x \mapsto (\tau_j \tau_i^{-1})_x$$

is a morphism.

Then we shall say that $\{(U_i, \tau_i)\}$ is a **trivializing covering** for π (or for E by abuse of language), and that $\{\tau_i\}$ are its **trivializing** maps. If $x \in U_i$, we say that τ_i (or U_i) trivializes at x. Two trivializing coverings for π are said to be **VB-equivalent** if taken together they also satisfy conditions **VB 2**, **VB 3**. An equivalence class of such trivializing coverings is said to determine a structure of **vector bundle** on π (or on E by abuse of language). We say that E is the **total space** of the bundle, and that X is its **base space**. If we wish to be very functorial, we shall write E_π and X_π for these spaces respectively. The fiber $\pi^{-1}(x)$ is also denoted by E_x or π_x. We also say that the vector bundle has **fiber E**, or is **modeled on E**. Note that from **VB 2**, the fiber $\pi^{-1}(x)$ above each point $x \in X$ can be given a structure of vector space, simply by transporting the vector space structure of $\mathbf{E}$ to $\pi^{-1}(x)$ via τ_{ix}. Condition **VB 2** insures that using two different trivializing maps τ_{ix} or τ_{jx} will give the same structure of vector space (with equivalent norms, of course not the same norms).

Conversely, we could replace **VB 2** by a similar condition as follows.

VB 2′. *On each fiber $\pi^{-1}(x)$ we are given a structure of vector space, and for $x \in U_i$, the trivializing map*

$$\tau_{ix}: \pi^{-1}(x) = E_x \to \mathbf{E}$$

is a linear isomorphism.

Then it follows that $\tau_{jx} \circ \tau_{ix}^{-1}: \mathbf{E} \to \mathbf{E}$ is a linear isomorphism for each pair of open sets U_i, U_j and $x \in U_i \cap U_j$.

Condition **VB 3** is implied by **VB 2**.

Proposition 1.1. *Let $\mathbf{E}$, $\mathbf{F}$ be vector spaces. Let U be open in some vector space. Let*

$$f: U \times \mathbf{E} \to \mathbf{F}$$

be a morphism such that for each $x \in U$, the map

$$f_x: \mathbf{E} \to \mathbf{F}$$

given by $f_x(v) = f(x, v)$ is a linear map. Then the map of U into $L(\mathbf{E}, \mathbf{F})$ given by $x \mapsto f_x$ is a morphism.

Proof. We can write $\mathbf{F} = \mathbf{R}_1 \times \cdots \times \mathbf{R}_n$ (n copies of $\mathbf{R}$). Using the fact that $L(\mathbf{E}, \mathbf{F}) = L(\mathbf{E}, \mathbf{R}_1) \times \cdots \times L(\mathbf{E}, \mathbf{R}_n)$, it will suffice to prove our assertion when $\mathbf{F} = \mathbf{R}$. Similarly, we can assume that $\mathbf{E} = \mathbf{R}$ also. But in that case, the function $f(x, v)$ can be written $g(x)v$ for some map $g: U \to \mathbf{R}$. Since f is a morphism, it follows that as a function of each argument x, v it is also a morphism. Putting $v = 1$ shows that g is a morphism and concludes the proof.

Returning to the general definition of a vector bundle, we call the maps

$$\tau_{jix} = \tau_{jx} \circ \tau_{ix}^{-1}$$

the **transition** maps associated with the covering. They satisfy what we call the **cocycle condition**

$$\tau_{kjx} \circ \tau_{jix} = \tau_{kix}.$$

In particular, $\tau_{iix} = \mathrm{id}$ and $\tau_{jix} = \tau_{ijx}^{-1}$.

As with manifolds, we can recover a vector bundle from a trivializing covering.

Proposition 1.2. *Let X be a manifold, and $\pi: E \to X$ a mapping from some set E into X. Let $\{U_i\}$ be an open covering of X, and for each i*

suppose that we are given a vector space **E** *and a bijection* (*commuting with the projection on U_i*),

$$\tau_i\colon \pi^{-1}(U_i) \to U_i \times \mathbf{E},$$

such that for each pair i, j and $x \in U_i \cap U_j$, the map $(\tau_j \tau_i^{-1})_x$ is a linear isomorphism, and condition **VB 3** *is satisfied as well as the cocycle condition. Then there exists a unique structure of manifold on E such that π is a morphism, such that τ_i is an isomorphism making π into a vector bundle, and $\{(U_i, \tau_i)\}$ into a trivialising covering.*

Proof. By Proposition 3.10 of Chapter I and our condition **VB 3**, we conclude that the map

$$\tau_j \tau_i^{-1}\colon \ (U_i \cap U_j) \times \mathbf{E} \to (U_i \cap U_j) \times \mathbf{E}$$

is a morphism, and in fact an isomorphism since it has an inverse. From the definition of atlases, we conclude that E has a unique manifold structure such that the τ_i are isomorphisms. Since π is obtained locally as a composite of morphisms (namely τ_i and the projections of $U_i \times \mathbf{E}$ on the first factor), it becomes a morphism. On each fiber $\pi^{-1}(x)$, we can transport the vector space structure of any **E** such that x lies in U_i, by means of τ_{ix}. The result is independent of the choice of U_i since $(\tau_j \tau_i^{-1})_x$ is a linear isomorphism. Our proposition is proved.

Remark. It is relatively rare that a vector bundle is **trivial**, i.e. VB-isomorphic to a product $X \times \mathbf{E}$. By definition, it is always trivial locally. In the finite dimensional case, say when E has dimension n, a trivialization is equivalent to the existence of sections $\xi_1, \ldots, \xi_n$ such that for each x, the vectors $\xi_1(x), \ldots, \xi_n(x)$ form a basis of E_x. Such a choice of sections is called a **frame** of the bundle, and is used especially with the tangent bundle, to be defined below.

The local representation of a vector bundle and the vector component of a morphism

For arbitrary vector bundles (and especially the tangent bundle to be defined below), we have a local representation of the bundle as a product in a chart. For many purposes, and especially the case of a morphism

$$f\colon Y \to E$$

of a manifold into the vector bundle, it is more convenient to use U to denote an open subset of a vector space, and to let $\varphi\colon U \to X$ be an isomorphism of U with an open subset of X over which E has a trivialization $\tau\colon \pi^{-1}(\varphi U) \to U \times \mathbf{E}$ called a **VB-chart**. Suppose V is an open subset of Y such that $f(V) \subset \pi^{-1}(\varphi U)$. We then have the commutative diagram:

$$
\begin{array}{ccccc}
V & \xrightarrow{\ f\ } & \pi^{-1}(\varphi U) & \xrightarrow{\ \tau\ } & U \times \mathbf{E} \\
 & & \downarrow & & \downarrow \\
 & & \varphi U & \xrightarrow{\ \varphi^{-1}\ } & U
\end{array}
$$

The composite $\tau \circ f$ is a morphism of V into $U \times \mathbf{E}$, which has two components

$$
\tau \circ f = (f_{U1},\ f_{U2})
$$

such that $f_{U1}\colon V \to U$ and $f_{U2}\colon V \to \mathbf{E}$. We call f_{U2} the **vector component of f in the vector bundle chart** $U \times \mathbf{E}$ over U. Sometimes to simplify the notation, we omit the subscript, and merely agree that $f_U = f_{U2}$ denotes this vector component; or to simplify the notation further, we may simply state that f itself denotes this vector component if a discussion takes place entirely in a chart. In this case, we say that $f = f_U$ **represents the morphism** in the vector bundle chart, or in the chart.

Vector bundle morphisms and pull backs

We now make the set of vector bundles into a category.

Let $\pi\colon E \to X$ and $\pi'\colon E' \to X'$ be two vector bundles. A **VB-morphism** $\pi \to \pi'$ consists of a pair of morphisms

$$
f_0\colon X \to X' \qquad \text{and} \qquad f\colon E \to E'
$$

satisfying the following conditions.

VB Mor 1. *The diagram*

$$
\begin{array}{ccc}
E & \xrightarrow{\ f\ } & E' \\
\pi \downarrow & & \downarrow \pi' \\
X & \xrightarrow[f_0]{} & X'
\end{array}
$$

is commutative, and the induced map for each $x \in X$

$$
f_x\colon E_x \to E'_{f(x)}
$$

is a linear map.

VB Mor 2. *For each $x_0 \in X$ there exist trivializing maps*

$$\tau: \pi^{-1}(U) \to U \times \mathbf{E}$$

and

$$\tau': \pi'^{-1}(U') \to U' \times \mathbf{E}'$$

at x_0 and $f(x_0)$ respectively, such that $f_0(U)$ is contained in U', and such that the map of U into $L(\mathbf{E}, \mathbf{E}')$ given by

$$x \mapsto \tau'_{f_0(x)} \circ f_x \circ \tau^{-1}$$

is a morphism.

As a matter of notation, we shall also use f to denote the VB-morphism, and thus write $f: \pi \to \pi'$. In most applications, f_0 is the identity. By Proposition 1.1, we observe that **VB Mor 2** is redundant.

The next proposition is the analogue of Proposition 1.2 for VB-morphisms.

Proposition 1.3. *Let π, π' be two vector bundles over manifolds X, X' respectively. Let $f_0: X \to X'$ be a morphism, and suppose that we are given for each $x \in X$ a continuous linear map*

$$f_x: \pi_x \to \pi'_{f_0(x)}$$

*such that, for each x_0, condition **VB Mor 2** is satisfied. Then the map f from π to π' defined by f_x on each fiber is a VB-morphism.*

Proof. One must first check that f is a morphism. This can be done under the assumption that π, π' are trivial, say equal to $U \times \mathbf{E}$ and $U' \times \mathbf{E}'$ (following the notation of **VB Mor 2**), with trivialising maps equal to the identity. Our map f is then given by

$$(x, v) \mapsto (f_0 x, f_x v).$$

Using Proposition 3.10 of Chapter I, we conclude that f is a morphism, and hence that (f_0, f) is a VB-morphism.

It is clear how to compose two VB-morphisms set theoretically. In fact, the composite of two VB-morphisms is a VB-morphism. There is no problem verifying condition **VB Mor 1**, and for **VB Mor 2**, we look at the situation locally. We encounter a commutative diagram of the following type:

$$\pi^{-1}(U) \xrightarrow{\ f\ } \pi'^{-1}(U') \xrightarrow{\ g\ } \pi''^{-1}(U'')$$

$$\left\downarrow{\scriptstyle \tau} \qquad\qquad \left\downarrow{\scriptstyle \tau'} \qquad\qquad \left\downarrow{\scriptstyle \tau''}$$

$$U \times \mathbf{E} \longrightarrow U' \times \mathbf{E'} \longrightarrow U'' \times \mathbf{E''}$$

and use Proposition 3.10 of Chapter I, to show that $g \circ f$ is a VB-morphism.

We therefore have a category, denoted by VB or VB^p, if we need to specify explicitly the order of differentiability.

The vector bundles over X from a subcategory $\mathrm{VB}(X) = \mathrm{VB}^p(X)$ (taking those VB-morphisms for which the map f_0 is the identity).

A morphism from one vector bundle into another can be given locally. More precisely, suppose that U is an open subset of X and $\pi \colon E \to X$ a vector bundle over X. Let $E_U = \pi^{-1}(U)$ and

$$\pi_U = \pi \,|\, E_U$$

be the restriction of π to E_U. Then π_U is a vector bundle over U. Let $\{U_i\}$ be an open covering of the manifold X and let π, π' be two vector bundles over X. Suppose, given a VB-morphism

$$f_i \colon \pi_{U_i} \to \pi'_{U_i}$$

for each i, such that f_i and f_j agree over $U_i \cap U_j$ for each pair of indices i, j. Then there exists a unique VB-morphism $f \colon \pi \to \pi'$ which agrees with f_i on each U_i. The proof is trivial, but the remark will be used frequently in the sequel.

Using the discussion at the end of Chapter II, §2 and Proposition 2.7 of that chapter, we get immediately:

Proposition 1.4. *Let $\pi \colon E \to Y$ be a vector bundle, and $f \colon X \to Y$ a morphism. Then*

$$f^*(\pi) \colon f^*(E) \to X$$

is a vector bundle called the **pull-back**, *and the pair $\left(f, \pi^*(f)\right)$ is a VB-morphism*

$$f^*(E) \xrightarrow{\ \pi^*(f)\ } E$$

$$\left\downarrow{\scriptstyle f^*(\pi)} \qquad\qquad \left\downarrow{\scriptstyle \pi}$$

$$X \xrightarrow[\ f\ ]{} Y$$

In Proposition 1.4, we could take f to be the inclusion of a sub-manifold. In that case, the pull-back is merely the restriction. As with open sets, we can then use the usual notation:

$$E_X = \pi^{-1}(X) \qquad \text{and} \qquad \pi_X = \pi \,|\, E_X.$$

Thus $\pi_X = f^*(\pi)$ in that case.

If X happens to be a point y of Y, then we have the constant map

$$\pi_y \colon E_y \to y$$

which will sometimes be identified with E_y.

If we identify each fiber $(f^*E)_x$ with $E_{f(x)}$ itself (a harmless identification since an element of the fiber at x is simply a pair (x, e) with e in $E_{f(x)}$), then we can describe the pull-back f^* of a vector bundle $\pi \colon E \to Y$ as follows. It is a vector bundle $f^*\pi \colon f^*E \to X$ satisfying the following properties:

PB 1. *For each* $x \in X$, *we have* $(f^*E)_x = E_{f(x)}$.

PB 2. *We have a commutative diagram*

$$
\begin{array}{ccc}
f^*(E) & \longrightarrow & E \\
{\scriptstyle f^*(\pi)}\big\downarrow & & \big\downarrow{\scriptstyle \pi} \\
X & \underset{f}{\longrightarrow} & Y
\end{array}
$$

the top horizontal map being the identity on each fiber.

PB 3. *If* E *is trivial, equal to* $Y \times \mathbf{E}$, *then* $f^*E = X \times \mathbf{E}$ *and* $f^*\pi$ *is the projection.*

PB 4. *If* V *is an open subset of* Y *and* $U = f^{-1}(V)$, *then*

$$f^*(E_V) = (f^*E)_U,$$

and we have a commutative diagram:

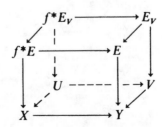

III, §2. THE TANGENT BUNDLE

Let X be a manifold of class C^p with $p \geq 1$. We shall define a functor T from the category of such manifolds into the category of vector bundles of class C^{p-1}.

For each manifold X we let $T(X)$ be the disjoint union of the tangent spaces $T_x(X)$. We have a natural projection

$$\pi \colon T(X) \to X$$

mapping $T_x(X)$ on x. We must make this into a vector bundle. If (U, φ) is a chart of X such that φU is open in the vector space $\mathbf{E}$, then from the definition of the tangent vectors as equivalence classes of triples (U, φ, v) we get immediately a bijection

$$\tau_U \colon \pi^{-1}(U) = T(U) \to U \times \mathbf{E}$$

which commutes with the projection on U, that is such that

$$\pi^{-1}(U) \xrightarrow{\ \tau_U\ } U \times \mathbf{E}$$
$$\searrow \qquad \swarrow$$
$$U$$

is commutative. Furthermore, if (U_i, φ_i) and (U_j, φ_j) are two charts, and if we denote by φ_{ji} the map $\varphi_j \varphi_i^{-1}$ (defined on $\varphi_i(U_i \cap U_j)$), then we obtain a transition mapping

$$\tau_{ji} = (\tau_j \tau_i^{-1}) \colon \varphi_i(U_i \cap U_j) \times \mathbf{E} \to \varphi_j(U_i \cap U_j) \times \mathbf{E}$$

by the formula

$$\tau_{ji}(x, v) = (\varphi_{ji}x, D\varphi_{ji}(x) \cdot v)$$

for $x \in U_i \cap U_j$ and $v \in \mathbf{E}$. Since the derivative $D\varphi_{ji} = \varphi'_{ji}$ is of class C^{p-1} and is an isomorphism at x, we see immediately that all the conditions of Proposition 1.2 are verified (using Proposition 3.10 of Chapter I), thereby making $T(X)$ into a vector bundle of class C^{p-1}.

We see that the above construction can also be expressed as follows. If the manifold X is glued together from open sets $\{U_i\}$ in vector spaces by means of transition mappings $\{\varphi_{ij}\}$, then we can glue together products $U_i \times \mathbf{E}$ by means of transition mappings $(\varphi_{ij}, D\varphi_{ij})$ where the derivative $D\varphi_{ij}$ can be viewed as a function of two variables (x, v). Thus locally, for open subsets U of vector spaces, the tangent bundle can be identified with the product $U \times \mathbf{E}$. The reader will note that our definition coincides with the oldest definition employed by geometers, our tangent vectors being vectors which transform according to a certain rule (namely the derivative).

If $f: X \to X'$ is a C^p-morphism, we can define

$$Tf: T(X) \to T(X')$$

to be simply $T_x f$ on each fiber $T_x(X)$. In order to verify that Tf is a VB-morphism (of class C^{p-1}), it suffices to look at the situation locally, i.e. we may assume that X and X' are open in vector spaces $\mathbf{E}$, $\mathbf{E}'$, and that $T_x f = f'(x)$ is simply the derivative. Then the map Tf is given by

$$Tf(x, v) = \big(f(x), f'(x)v\big)$$

for $x \in X$ and $v \in \mathbf{E}$. Since f' is of class C^{p-1} by definition, we can apply Proposition 3.10 of Chapter I to conclude that Tf is also of class C^{p-1}. The functoriality property is trivially satisfied, and we have therefore defined the functor T as promised.

It will sometimes be notationally convenient to write f_* instead of Tf for the induced map, which is also called the **tangent map**. The bundle $T(X)$ is called the **tangent bundle** of X.

Remark. The above definition of the tangent bundle fits with Steenrod's point of view [Ste 51]. I don't understand why many differential geometers have systematically rejected this point of view, when they take the definition of a tangent vector as a differential operator.

III, §3. EXACT SEQUENCES OF BUNDLES

Let X be a manifold. Let $\pi': E' \to X$ and $\pi: E \to X$ be two vector bundles over X. Let $f: \pi' \to \pi$ be a VB-morphism. We shall say that the sequence

$$0 \to \pi' \xrightarrow{f} \pi$$

is **exact** if there exists a covering of X by open sets and for each open set U in this covering there exist trivializations

$$\tau': E'_U \to U \times \mathbf{E}' \qquad \text{and} \qquad \tau: E_U \to U \times \mathbf{E}$$

such that $\mathbf{E}$ can be written as a product $\mathbf{E} = \mathbf{E}' \times \mathbf{F}$, making the following diagram commutative:

$$
\begin{array}{ccc}
E'_U & \xrightarrow{\ f\ } & E_U \\
{\scriptstyle \tau'}\downarrow & & \downarrow{\scriptstyle \tau} \\
U \times \mathbf{E}' & \longrightarrow & U \times \mathbf{E}' \times \mathbf{F}
\end{array}
$$

(The bottom map is the natural one: Identity on U and the injection of $\mathbf{E}'$ on $\mathbf{E}' \times 0$.)

Let $\pi_1: E_1 \to X$ be another vector bundle, and let $g: \pi_1 \to \pi$ be a VB-morphism such that $g(E_1)$ is contained in $f(E')$. Since f establishes a bijection between E' and its image $f(E')$ in E, it follows that there exists a unique map $g_1: E_1 \to E'$ such that $g = f \circ g_1$. We contend that g_1 is a VB-morphism. Indeed, to prove this we can work locally, and in view of the definition, over an open set U as above, we can write

$$g_1 = \tau'^{-1} \circ \mathrm{pr} \circ \tau \circ g$$

where pr is the projection of $U \times \mathbf{E}' \times \mathbf{F}$ on $U \times \mathbf{E}'$. All the maps on the right-hand side of our equality are VB-morphisms; this proves our contention.

Let $\pi: E \to X$ be a vector bundle. A subset S of E will be called a **subbundle** if there exists an exact sequence $0 \to \pi' \to \pi$, also written

$$0 \to E' \xrightarrow{f} E,$$

such that $f(E') = S$. This gives S the structure of a vector bundle, and the previous remarks show that it is unique. In fact, given another exact sequence

$$0 \to E_1 \xrightarrow{g} E$$

such that $g(E_1) = S$, the natural map $f^{-1}g$ from E_1 to E' is a VB-isomorphism.

Let us denote by E/E' the union of all factor spaces E_x/E'_x. If we are dealing with an exact sequence as above, then we can give E/E' the structure of a vector bundle. We proceed as follows. Let $\{U_i\}$ be our covering, with trivialising maps τ'_i and τ_i. We can define for each i a bijection

$$\pi''_i: E_{U_i}/E'_{U_i} \to U_i \times \mathbf{F}$$

obtained in a natural way from the above commutative diagram. (Without loss of generality, we can assume that the vector spaces $\mathbf{E}'$, $\mathbf{F}$ are constant for all i.) We have to prove that these bijections satisfy the conditions of Proposition 1.2.

Without loss of generality, we may assume that f is an inclusion (of the total space E' into E). For each pair i, j and $x \in U_i \cap U_j$, the toplinear automorphism $(\tau_j \tau_i^{-1})_x$ is represented by a matrix

$$\begin{pmatrix} h_{11}(x) & h_{12}(x) \\ h_{21}(x) & h_{22}(x) \end{pmatrix}$$

operating on the right on a vector $(v, w) \in \mathbf{E}' \times \mathbf{F}$. The map $(\tau''_j \tau''^{-1}_i)_x$ on $\mathbf{F}$ is induced by this matrix. Since $\mathbf{E}' = \mathbf{E}' \times 0$ has to be carried into itself by the matrix, we have $h_{12}(x) = 0$. Furthermore, since $(\tau_j \tau_i^{-1})_x$ has an inverse, equal to $(\tau_i \tau_j^{-1})_x$, it follows that $h_{22}(x)$ is a toplinear automorphism of $\mathbf{F}$, and represents $(\tau_j'' \tau_i''^{-1})_x$. Therefore condition **VB 3** is satisfied, and E/E' is a vector bundle.

The canonical map

$$E_U \to E_U/E_U'$$

is a morphism since it can be expressed in terms of τ, the projection, and τ''^{-1}. Consequently, we obtain a VB-morphism

$$g \colon \pi \to \pi''$$

in the canonical way (on the total spaces, it is the quotient mapping of E on E/E'). We shall call π'' the **factor bundle**.

Our map g satisfies the usual universal mapping property of a cokernel. Indeed, suppose that

$$\psi \colon E \to G$$

is a VB-morphism such that $\psi \circ f = 0$ (i.e. $\psi_x \circ f_x = 0$ on each fiber E_x'). We can then define set theoretically a canonical map

$$\psi_* \colon E/E' \to G,$$

and we must prove that it is a VB-morphism. This can be done locally. Using the above notation, we may assume that $E = U \times \mathbf{E}' \times \mathbf{F}$ and that g is the projection. In that case, ψ_* is simply the canonical injection of $U \times \mathbf{F}$ in $U \times \mathbf{E}' \times \mathbf{F}$ followed by ψ, and is therefore a VB-morphism.

We shall therefore call g the **cokernel** of f.

Dually, let $g \colon \pi \to \pi''$ be a given VB-morphism. We shall say that the sequence

$$\pi \xrightarrow{g} \pi'' \to 0$$

is **exact** if g is surjective, and if there exists a covering of X by open sets, and for each open set U in this covering there exist spaces $\mathbf{E}'$, $\mathbf{F}$ and trivializations

$$\tau \colon E_U \to U \times \mathbf{E}' \times \mathbf{F} \qquad \text{and} \qquad \tau'' \colon \mathbf{E}_U'' \to \mathbf{F}$$

making the following diagram commutative:

$$
\begin{array}{ccc}
E_U & \xrightarrow{\;\;g\;\;} & E_U'' \\
\tau \downarrow & & \downarrow \tau'' \\
U \times \mathbf{E}' \times \mathbf{F} & \longrightarrow & U \times \mathbf{F}
\end{array}
$$

(The bottom map is the natural one: Identity on U and the projection of $\mathbf{E}' \times \mathbf{F}$ on $\mathbf{F}$.)

In the same way as before, one sees that the "kernel" of g, that is, the union of the kernels E'_x of each g_x, can be given a structure of vector bundle. This union E' will be called the **kernel** of g, and satisfies the usual universal mapping property.

Proposition 3.1. *Let X be a manifold and let*

$$f \colon \pi' \to \pi$$

be a VB-morphism of vector bundles over X. Assume that, for each $x \in X$, the continuous linear map

$$f_x \colon E'_x \to E_x$$

is injective. Then the sequence

$$0 \to \pi' \xrightarrow{f} \pi$$

is exact.

Proof. We can assume that X is connected and that the fibers of E' and E are constant, say equal to the vector spaces $\mathbf{E}'$ and $\mathbf{E}$. Let $a \in X$. Corresponding to the splitting of f_a we know that we have a product decomposition $\mathbf{E} = \mathbf{E}' \times \mathbf{F}$ and that there exists an open set U of X containing a, together with trivializing maps

$$\tau \colon \pi^{-1}(U) \to U \times \mathbf{E} \qquad \text{and} \qquad \tau' \colon \pi'^{-1}(U) \to U \times \mathbf{E}'$$

such that the composite map

$$\mathbf{E}' \xrightarrow{\tau_a'^{-1}} E'_a \xrightarrow{f_a} E_a \xrightarrow{\tau_a} \mathbf{E}' \times \mathbf{F}$$

maps $\mathbf{E}'$ on $\mathbf{E}' \times 0$.

For any point x in U, we have a map

$$(\tau f \tau'^{-1})_x \colon \mathbf{E}' \to \mathbf{E}' \times \mathbf{F},$$

which can be represented by a pair of continuous linear maps

$$\big(h_{11}(x), h_{21}(x)\big).$$

We define

$$h(x) \colon \mathbf{E}' \times \mathbf{F} \to \mathbf{E}' \times \mathbf{F}$$

by the matrix

$$\begin{pmatrix} h_{11}(x) & 0 \\ h_{21}(x) & \mathrm{id} \end{pmatrix},$$

operating on the right on a vector $(v, w) \in \mathbf{E}' \times \mathbf{F}$. Then $h(x)$ restricted to $\mathbf{E}' \times 0$ has the same action as $(\tau f \tau'^{-1})_x$.

The map $x \mapsto h(x)$ is a morphism of U into $L(\mathbf{E}, \mathbf{E})$ and since it is continuous, it follows that for U small enough around our fixed point a, it maps U into the group of linear automorphisms of $\mathbf{E}$. This proves our proposition.

Dually to Proposition 3.1, we have:

Proposition 3.2. *Let X be a manifold and let*

$$g: \pi \to \pi''$$

be a VB-morphism of vector bundles over X. Assume that for each $x \in X$, the continuous linear map

$$g_x: E_x \to E''_x$$

is surjective. Then the sequence

$$\pi \xrightarrow{g} \pi'' \to 0$$

is exact.

Proof. It is dual to the preceding one and we leave it to the reader.

In general, a sequence of VB-morphisms

$$0 \to \pi' \xrightarrow{f} \pi \xrightarrow{g} \pi'' \to 0$$

is said to be **exact** if both ends are exact, and if the image of f is equal to the kernel of g.

There is an important example of exact sequence. Let $f: X \to Y$ be an immersion. By the universal mapping property of pull backs, we have a canonical VB-morphism

$$T^*f: T(X) \to f^*T(Y)$$

of $T(X)$ into the pull back over X of the tangent bundle of Y. Furthermore, from the manner in which the pull back is obtained locally by taking products, and the definition of an immersion, one sees that the sequence

$$0 \to T(X) \xrightarrow{T^*f} f^*T(Y)$$

is exact. The factor bundle

$$f^*T(Y)/\mathrm{Im}(T^*f)$$

is called the **normal** bundle of f. It is denoted by $N(f)$, and its total space by $N_f(X)$ if we wish to distinguish between the two. We sometimes identify $T(X)$ with its image under T^*f and write

$$N(f) = f^*T(Y)/T(X).$$

Dually, let $f: X \to Y$ be a submersion. Then we have an exact sequence

$$T(X) \xrightarrow{T^*f} f^*T(Y) \to 0$$

whose kernel could be called the **subbundle** of f, or the **bundle along the fiber**.

There is an interesting case where we can describe the kernel more precisely. Let

$$\pi: E \to X$$

be a vector bundle. Then we can form the pull back of E over itself, that is, π^*E, and we contend that we have an exact sequence

$$0 \to \pi^*E \to T(E) \to \pi^*T(X) \to 0.$$

To define the map on the left, we look at the subbundle of π more closely. For each $x \in X$ we have an inclusion

$$E_x \to E,$$

whence a natural injection

$$T(E_x) \to T(E).$$

The local product structure of a bundle shows that the union of the $T(E_x)$ as x ranges over X gives the subbundle set theoretically. On the other hand, the total space of π^*E consists of pairs of vectors (v, w) lying over the same base point x, that is, the fiber at x of π^*E is simply $E_x \times E_x$. Since $T(E_x)$ has a natural identification with $E_x \times E_x$, we get for each x a bijection

$$(\pi^*E)_x \to T(E_x)$$

which defines our map from π^*E to $T(E)$. Considering the map locally in terms of the local product structure shows at once that it gives a VB-isomorphism between π^*E and the subbundle of π, as desired.

III, §4. OPERATIONS ON VECTOR BUNDLES

We consider a functor

$$(\mathbf{E}, \mathbf{F}) \mapsto \lambda(\mathbf{E}, \mathbf{F})$$

in, say, two variables, which is, say, contravariant in the first and co-variant in the second. (Everything we shall do extends in the obvious manner to functors of several variables.

Example. We took a functor in two variables for definiteness, and to illustrate both variances. However, we could consider a functor in one or more than two variables. For instance, let us consider the functor

$$\mathbf{E} \mapsto L(\mathbf{E}, \mathbf{R}) = L(\mathbf{E}) = \mathbf{E}^\vee,$$

which we call the **dual**. It is a contravariant functor in one variable. On the other hand, the functor

$$\mathbf{E} \mapsto L_a^r(\mathbf{E}, \mathbf{F})$$

of continuous multilinear maps of $\mathbf{E} \times \cdots \times \mathbf{E}$ into a vector space $\mathbf{F}$ is contravariant in $\mathbf{E}$ and covariant in $\mathbf{F}$. The functor $\mathbf{E} \mapsto L_a^r(\mathbf{E}, \mathbf{R})$ gives rise later to what we call differential forms. We shall treat such forms systematically in Chapter V, §3.

Let $f \colon \mathbf{E}' \to \mathbf{E}$ and $g \colon \mathbf{F} \to \mathbf{F}'$ be two linear maps. By definition, we have a map

$$L(\mathbf{E}', \mathbf{E}) \times L(\mathbf{F}, \mathbf{F}') \to L\big(\lambda(\mathbf{E}, \mathbf{F}), \lambda(\mathbf{E}', \mathbf{F}')\big),$$

assigning $\lambda(f, g)$ to (f, g).

We shall say that λ is of **class** C^p if the following condition is satisfied. Give a manifold U, and two morphisms

$$\varphi \colon U \to L(\mathbf{E}', \mathbf{E}) \qquad \text{and} \qquad \psi \colon U \to L(\mathbf{F}, \mathbf{F}'),$$

then the composite

$$U \to L(\mathbf{E}', \mathbf{E}) \times L(\mathbf{F}, \mathbf{F}') \to L\big(\lambda(\mathbf{E}, \mathbf{F}), \lambda(\mathbf{E}', \mathbf{F}')\big)$$

is also a morphism. (One could also say that λ is **differentiable**.)

Theorem 4.1. *Let λ be a functor as above, of class C^p, $p \geqq 0$. Then for each manifold X, there exists a functor λ_X, on vector bundles (of class C^p)*

$$\lambda_X \colon \mathrm{VB}(X) \times \mathrm{VB}(X) \to \mathrm{VB}(X)$$

satisfying the following properties. For any bundles α, β *in* **VB**(X) *and* **VB***-morphisms*

$$f\colon \alpha' \to \alpha \qquad and \qquad g\colon \beta \to \beta'$$

and for each $x \in X$, *we have*:

OP 1. $\lambda_X(\alpha, \beta)_x = \lambda(\alpha_x, \beta_x)$.

OP 2. $\lambda_X(f, g)_x = \lambda(f_x, g_x)$.

OP 3. *If* α *is the trivial bundle* $X \times \mathbf{E}$ *and* β *the trivial bundle* $X \times \mathbf{F}$, *then* $\lambda_X(\alpha, \beta)$ *is the trivial bundle* $X \times \lambda(\mathbf{E}, \mathbf{F})$.

OP 4. *If* $h\colon Y \to X$ *is a* C^p-*morphism, then*

$$\lambda_Y^*(h^*\alpha, h^*\beta) = h^*\lambda_X(\alpha, \beta).$$

Proof. We may assume that X is connected, so that all the fibers are linearly isomorphic to a fixed space. For each open subset U of X we let the total space $\lambda_U(E_\alpha, E_\beta)$ of $\lambda_U(\alpha, \beta)$ be the union of the sets

$$\{x\} \times \lambda(\alpha_x, \beta_x)$$

(identified harmlessly throughout with $\lambda(\alpha_x, \beta_x)$), as x ranges over U. We can find a covering $\{U_i\}$ of X with trivializing maps $\{\tau_i\}$ for α, and $\{\sigma_i\}$ for β,

$$\tau_i\colon \alpha^{-1}(U_i) \to U_i \times \mathbf{E},$$

$$\sigma_i\colon \beta^{-1}(U_i) \to U_i \times \mathbf{F}.$$

We have a bijection

$$\lambda(\tau_i^{-1}, \sigma_i)\colon \lambda_{U_i}(E_\alpha, E_\beta) \to U_i \times \lambda(\mathbf{E}, \mathbf{F})$$

obtained by taking on each fiber the map

$$\lambda(\tau_{ix}^{-1}, \sigma_{ix})\colon \lambda(\alpha_x, \beta_x) \to \lambda(\mathbf{E}, \mathbf{F}).$$

We must verify that **VB 3** is satisfied. This means looking at the map

$$x \to \lambda(\tau_{jx}^{-1}, \sigma_{jx}) \circ \lambda(\tau_{ix}^{-1}, \sigma_{ix})^{-1}.$$

The expression on the right is equal to

$$\lambda(\tau_{ix}\tau_{jx}^{-1}, \sigma_{jx}\sigma_{ix}^{-1}).$$

Since λ is a functor of class C^p, we see that we get a map

$$U_i \cap U_j \to L\big(\lambda(\mathbf{E}, \mathbf{F}), \lambda(\mathbf{E}, \mathbf{F})\big)$$

which is a C^p-morphism. Furthermore, since λ is a functor, the transition mappings are in fact linear isomorphisms, and **VB 2**, **VB 3** are proved.

The proof of the analogous statement for $\lambda_X(f, g)$, to the effect that it is a VB-morphism, proceeds in an analogous way, again using the hypothesis that λ is of class C^p. Condition **OP 3** is obviously satisfied, and **OP 4** follows by localizing. This proves our theorem.

The next theorem gives us the uniqueness of the operation λ_X.

Theorem 4.2. *If μ is another functor of class C^p with the same variance as λ, and if we have a natural transformation of functors $t: \lambda \to \mu$, then for each X, the mapping*

$$t_X: \lambda_X \to \mu_X,$$

defined on each fiber by the map

$$t(\alpha_x, \beta_x): \lambda(\alpha_x, \beta_x) \to \mu(\alpha_x, \beta_x),$$

is a natural transformation of functors (in the VB-category).

Proof. For simplicity of notation, assume that λ and μ are both functors of one variable, and both covariant. For each open set $U = U_i$ of a trivializing covering for β, we have a commutative diagram:

The vertical maps are trivializing VB-isomorphism, and the top horizontal map is a VB-morphism. Hence t_U is a VB-morphism, and our assertion is proved.

In particular, for $\lambda = \mu$ and $t = $ id we get the uniqueness of our functor λ_X.

(In the proof of Theorem 4.2, we do not use again explicitly the hypotheses that λ, μ are differentiable.)

In practice, we omit the subscript X on λ, and write λ for the functor on vector bundles.

Examples. Let $\pi\colon E \to X$ be a vector bundle. We take λ to be the dual, that is $\mathbf{E} \mapsto \mathbf{E}^\vee = L(\mathbf{E}, \mathbf{R})$. Then $\lambda(E)$ is denoted by $E^\vee$, and is called the **dual bundle**. The fiber at each point $x \in X$ is the dual space $E_x^\vee$. The dual bundle of the tangent bundle is called the **cotangent bundle** $T^\vee X$.

Similarly, instead of taking $L(E)$, we could take $L_a^r(E)$ to be the bundle of alternating multilinear forms on E. The fiber at each point is the space $L_a^r(E_x)$ consisting of all r-multilinear alternating continuous functions on E_x. When $E = TX$ is the tangent bundle, the sections of $L_a^r(TX)$ are called **differential forms** of degree r. Thus a 1-form is a section of $E^\vee$. Differential forms will be treated later in detail.

For another type of operation, we have the **direct sum** (also called the **Whitney sum**) of two bundles α, β over X. It is denoted by $\alpha \oplus \beta$, and the fiber at a point x is

$$(\alpha \oplus \beta)_x = \alpha_x \oplus \beta_x.$$

Of course, the finite direct sum of vector spaces can be identified with their finite direct products, but we write the above operation as a direct sum in order not to confuse it with the following direct product.

Let $\alpha\colon E_\alpha \to X$ and $\beta\colon E_\beta \to Y$ be two vector bundles in $\mathrm{VB}(X)$ and $\mathrm{VB}(Y)$ respectively. Then the map

$$\alpha \times \beta\colon E_\alpha \times E_\beta \to X \times Y$$

is a vector bundle, and it is this operation which we call the **direct product** of α and β.

Let X be a manifold, and λ a functor of class C^p with $p \geq 1$. The **tensor bundle** of type λ over X is defined to be $\lambda_X\big(T(X)\big)$, also denoted by $\lambda T(X)$ or $T_\lambda(X)$. The sections of this bundle are called **tensor fields** of type λ, and the set of such sections is denoted by $\Gamma_\lambda(X)$. Suppose that we have a trivialization of $T(X)$, say

$$T(X) = X \times \mathbf{E}.$$

Then $T_\lambda(X) = X \times \lambda(\mathbf{E})$. A section of $T_\lambda(X)$ in this representation is completely described by the projection on the second factor, which is a morphism

$$f\colon X \to \lambda(\mathbf{E}).$$

We shall call it the **local representation** of the tensor field (in the given trivialization). If ξ is the tensor field having f as its local representation, then

$$\xi(x) = \big(x, f(x)\big).$$

Let $f: X \to Y$ be a morphism of class C^p ($p \geq 1$). Let ω be a tensor field of type L^r over Y, which could also be called a **multilinear tensor field**. For each $y \in Y$, $\omega(y)$ (also written ω_y) is a multilinear function on $T_y(Y)$:

$$\omega_y: T_y \times \cdots \times T_y \to \mathbf{R}.$$

For each $x \in X$, we can define a multilinear map

$$f_x^*(\omega): T_x \times \cdots \times T_x \to \mathbf{R}$$

by the composition of maps $(T_x f)^r$ and $\omega_{f(x)}$:

$$T_x \times \cdots \times T_x \to T_{f(x)} \times \cdots \times T_{f(x)} \to \mathbf{R}.$$

We contend that the map $x \mapsto f_x^*(\omega)$ is a tensor field over X, of the same type as ω. To prove this, we may work with local representation. Thus we can assume that we work with a morphism

$$f: U \to V$$

of one open set in a Banach space into another, and that

$$\omega: V \to L^r(\mathbf{F})$$

is a morphism, V being open in $\mathbf{F}$. If U is open in $\mathbf{E}$, then $f^*(\omega)$ (now denoting a local representation) becomes a mapping of U into $L^r(\mathbf{E})$, given by the formula

$$f_x^*(\omega) = L^r(f'(x)) \cdot \omega(f(x)).$$

Since $L^r: L(\mathbf{E}, \mathbf{F}) \to L(L^r(\mathbf{F}), L^r(\mathbf{E}))$ is of class C^∞, it follows that $f^*(\omega)$ is a morphism of the same class as ω. This proves what we want.

Of course, the same argument is valid for the other functors L_s^r and L_a^r (symmetric and alternating multilinear maps). Special cases will be considered in later chapters. If λ denotes any one of our three functors, then we see that we have obtained a mapping (which is in fact linear)

$$f^*: \Gamma_\lambda(Y) \to \Gamma_\lambda(X)$$

which is clearly functorial in f. We use the notation f^* instead of the more correct (but clumsy) notation f_λ or $\Gamma_\lambda(f)$. No confusion will arise from this.

III, §5. SPLITTING OF VECTOR BUNDLES

The next proposition expresses the fact that the VB-morphisms of one bundle into another (over a fixed morhism) form a module over the ring of functions.

Proposition 5.1. *Let X, Y be manifolds and $f_0: X \to Y$ a morphism. Let α, β be vector bundles over X, Y respectively, and let f, $g: \alpha \to \beta$ be two VB-morphisms over f_0. Then the map $f + g$ defined by the formula*

$$(f + g)_x = f_x + g_x$$

is also a VB-morphism. Furthermore, if $\psi: Y \to \mathbf{R}$ is a function on Y, then the map ψf defined by

$$(\psi f)_x = \psi(f_0(x)) f_x$$

is also a VB-morphism.

Proof. Both assertions are immediate consequences of Proposition 3.10 of Chapter I.

We shall consider mostly the situation where $X = Y$ and f_0 is the identity, and will use it, together with partitions of unity, to glue VB-morphisms together.

Let α, β be vector bundles over X and let $\{(U_i, \psi_i)\}$ be a partition of unity on X. Suppose given for each U_i a VB-morphism

$$f_i: \alpha|U_i \to \beta|U_i.$$

Each one of the maps $\psi_i f_i$ (defined as in Proposition 5.1) is a VB-morphism. Furthermore, we can extend $\psi_i f_i$ to a VB-morphism of α into β simply by putting

$$(\psi_i f_i)_x = 0$$

for all $x \notin U_i$. If we now define

$$f: \alpha \to \beta$$

by the formula

$$f_x(v) = \sum \psi_i(x) f_{ix}(v)$$

for all pairs (x, v) with $v \in \alpha_x$, then the sum is actually finite, at each ponit x, and again by Proposition 5.1, we see that f is a VB-morphism. We observe that if each f_i is the identity, then $f = \sum \psi_i f_i$ is also the identity.

Proposition 5.2. *Let X be a manifold admitting partitions of unity. Let $0 \to \alpha \xrightarrow{f} \beta$ be an exact sequence of vector bundles over X. Then there exists a surjective VB-morphism $g \colon \beta \to \alpha$ whose kernel splits at each point, such that $g \circ f = \mathrm{id}$.*

Proof. By the definition of exact sequence, there exists a partition of unity $\{(U_i, \psi_i)\}$ on X such that for each i, we can split the sequence over U_i. In other words, there exists for each i a VB-morphism

$$g_i \colon \beta|U_i \to \alpha|U_i$$

which is surjective, whose kernel splits, and such that $g_i \circ f_i = \mathrm{id}_i$. We let $g = \sum \psi_i g_i$. Then g is a VB-morphism of β into α by what we have just seen, and

$$g \circ f = \sum \psi_i g_i f_i = \mathrm{id}.$$

It is trivial that g is surjective because $g \circ f = \mathrm{id}$. The kernel of g_x splits at each point x because it has a closed complement, namely $f_x \alpha_x$. This concludes the proof.

If γ is the kernel of β, then we have $\beta \approx \alpha \oplus \gamma$.

A vector bundle π over X will be said to be of **finite type** if there exists a finite trivialization for π (i.e. a trivialization $\{(U_i, \tau_i)\}$ such that i ranges over a finite set).

If k is an integer ≥ 1 and $\mathbf{E}$ a vector space, then we denote by $\mathbf{E}^k$ the direct product of $\mathbf{E}$ with itself k times.

Proposition 5.3. *Let X be a manifold admitting partitions of unity. Let π be a vector bundle of finite type in $\mathrm{VB}(X, \mathbf{E})$. Then there exists an integer $k > 0$ and a vector bundle α in $\mathrm{VB}(X, \mathbf{E}^k)$ such that $\pi \oplus \alpha$ is trivializable.*

Proof. We shall prove that there exists an exact sequence

$$0 \to \pi \xrightarrow{f} \beta$$

with $E_\beta = X \times \mathbf{E}^k$. Our theorem will follow from the preceding proposition.

Let $\{U_i, \tau_i\}$ be a finite trivialization of π with $i = 1, \ldots, k$. Let $\{(U_i, \psi_i)\}$ be a partition of unity. We define

$$f \colon E_\pi \to X \times \mathbf{E}^k$$

as follows. If $x \in X$ and v is in the fiber of E_π at x, then

$$f_x(v) = \big(x, \psi_1(x)\tau_1(v), \ldots, \psi_k(x)\tau_k(v)\big).$$

The expression on the right makes sense, because in case x does not lie in U_i then $\psi_i(x) = 0$ and we do not have to worry about the expression $\tau_i(v)$. If x lies in U_i, then $\tau_i(v)$ means $\tau_{ix}(v)$.

Given any point x, there exists some index i such that $\psi_i(x) > 0$ and hence f is injective. Furthermore, for this x and this index i, f_x maps E_x onto a closed subspace of $\mathbf{E}^k$, which admits a closed complement, namely

$$\mathbf{E} \times \cdots \times 0 \times \cdots \times \mathbf{E}$$

with 0 in the i-th place. This proves our proposition.

CHAPTER IV

Vector Fields and Differential Equations

In this chapter, we collect a number of results all of which make use of the notion of differential equation and solutions of differential equations.

Let X be a manifold. A vector field on X assigns to each point x of X a tangent vector, differentiably. (For the precise definition, see §2.) Given x_0 in X, it is then possible to construct a unique curve $\alpha(t)$ starting at x_0 (i.e. such that $\alpha(0) = x_0$) whose derivative at each point is the given vector. It is not always possible to make the curve depend on time t from $-\infty$ to $+\infty$, although it is possible if X is compact.

The structure of these curves presents a fruitful domain of investigation, from a number of points of view. For instance, one may ask for topological properties of the curves, that is those which are invariant under topological automorphisms of the manifold. (Is the curve a closed curve, is it a spiral, is it dense, etc.?) More generally, following standard procedures, one may ask for properties which are invariant under any given interesting group of automorphisms of X (discrete groups, Lie groups, algebraic groups, Riemannian automorphisms, ad lib.).

We do not go into these theories, each of which proceeds according to its own flavor. We give merely the elementary facts and definitions associated with vector fields, and some simple applications of the existence theorem for their curves.

Throughout this chapter, we assume all manifolds to be of class C^p with $p \geq 2$ from §2 on, and $p \geq 3$ from §3 on. This latter condition insures that the tangent bundle is of class C^{p-1} with $p - 1 \geq 1$ (or 2).

We shall deal with mappings of several variables, say $f(t, x, y)$, the first of which will be a real variable. We identify $D_1 f(t, x, y)$ with

$$\lim_{h \to 0} \frac{f(t + h, x, y) - f(t, x, y)}{h}.$$

IV, §1. EXISTENCE THEOREM FOR DIFFERENTIAL EQUATIONS

Let $\mathbf{E}$ be a normed vector space and U an open subset of $\mathbf{E}$. In this section we consider vector fields locally. The notion will be globalized later, and thus for the moment, we define (the local representation of) a **time-dependent vector field** on U to be a C^p-morphism $(p \geq 0)$

$$f: J \times U \to \mathbf{E},$$

where J is an open interval containing 0 in $\mathbf{R}$. We think of f as assigning to each point x in U a vector $f(t, x)$ in $\mathbf{E}$, depending on time t.

Let x_0 be a point of U. An **integral curve** for f with **initial condition** x_0 is a mapping of class C^r $(r \geq 1)$

$$\alpha: J_0 \to U$$

of an open subinterval of J containing 0, into U, such that $\alpha(0) = x_0$ and such that

$$\alpha'(t) = f(t, \alpha(t)).$$

Remark. Let $\alpha: J_0 \to U$ be a continuous map satisfying the condition

$$\alpha(t) = x_0 + \int_0^t f(u, \alpha(u)) \, du.$$

Then α is differentiable, and its derivative is $f(t, \alpha(t))$. Hence α is of class C^1. Furthermore, we can argue recursively, and conclude that if f is of class C^p, then so is α. Conversely, if α is an integral curve for f with initial condition x_0, then it obviously satisfies out integral relation.

Let

$$f: J \times U \to \mathbf{E}$$

be as above, and let x_0 be a point of U. By a **local flow** for f at x_0 we mean a mapping

$$\alpha: J_0 \times U_0 \to U$$

where J_0 is an open subinterval of J containing 0, and U_0 is an open subset of U containing x_0, such that for each x in U_0 the map

$$\alpha_x(t) = \alpha(t, x)$$

is an integral curve for f with initial condition x (i.e. such that $\alpha(0, x) = x$).

As a matter of notation, when we have a mapping with two arguments, say $\varphi(t, x)$, then we denote the separate mappings in each argument when the other is kept fixed by $\varphi_x(t)$ and $\varphi_t(x)$. The choice of letters will always prevent ambiguity.

We shall say that f satisfies a **Lipschitz condition** on U **uniformly with respect to** J if there exists a number $K > 0$ such that

$$|f(t, x) - f(t, y)| \leqq K|x - y|$$

for all x, y in U and t in J. We call K a **Lipschitz constant**. If f is of class C^1, it follows at once from the mean value theorem that f is Lipschitz on some open neighborhood $J_0 \times U_0$ of a given point $(0, x_0)$ of U, and that it is bounded on some such neighborhood.

We shall now prove that under a Lipschitz condition, local flows exist and are unique locally. In fact, we prove more, giving a uniformity property for such flows. If b is real > 0, then we denote by J_b the open interval $-b < t < b$.

Proposition 1.1. *Let J be an open interval of $\mathbf{R}$ containing 0, and U open in the normed vector space $\mathbf{E}$. Let x_0 be a point of U, and $a > 0$, $a < 1$ a real number such that the closed ball $\bar{B}_{3a}(x_0)$ lies in U. Assume that we have a continuous map*

$$f: J \times U \rightarrow \mathbf{E}$$

which is bounded by a constant $L \geqq 1$ on $J \times U$, and satisfies a Lipschitz condition on U uniformly with respect to J, with constant $K \geqq 1$. If $b < a/LK$, then for each x in $\bar{B}_a(x_0)$ there exists a unique flow

$$\alpha: J_b \times B_a(x_0) \rightarrow U.$$

If f is of class C^p $(p \geqq 1)$, then so is each integral curve α_x.

Proof. Let I_b be the closed interval $-b \leqq t \leqq b$, and let x be a fixed point in $\bar{B}_a(x_0)$. Let M be the set of continuous maps

$$\alpha: I_b \rightarrow \bar{B}_{2a}(x_0)$$

of the closed interval into the closed ball of center x_0 and radius $2a$, such that $\alpha(0) = x$. Then M is a complete metric space if we define as usual the distance between maps α, β to be the sup norm

$$\|\alpha - \beta\| = \sup_{t \in I_b} |\alpha(t) - \beta(t)|.$$

We shall now define a mapping

$$S: M \to M$$

of M into itself. For each α in M, we let $S\alpha$ be defined by

$$(S\alpha)(t) = x + \int_0^t f(u, \alpha(u))\, du.$$

Then $S\alpha$ is certainly continuous, we have $S\alpha(0) = x$, and the distance of any point on $S\alpha$ from x is bounded by the norm of the integral, which is bounded by

$$b \sup |f(u, y)| \leqq bL < a.$$

Thus $S\alpha$ lies in M.

We contend that our map S is a shrinking map. Indeed,

$$|S\alpha - S\beta| \leqq b \sup |f(u, \alpha(u)) - f(u, \beta(u))|$$
$$\leqq bK |\alpha - \beta|,$$

thereby proving our contention.

By the shrinking lemma (Chapter I, Lemma 5.1) our map has a unique fixed point α, and by definition, $\alpha(t)$ satisfies the desired integral relation. Our remark above concludes the proof.

Corollary 1.2. *The local flow α in Proposition 1.1 is continuous. Furthermore, the map $x \mapsto \alpha_x$ of $\bar{B}_a(x_0)$ into the space of curves is continuous, and in fact satisfies a Lipschitz condition.*

Proof. The second statement obviously implies the first. So fix x in $\bar{B}_a(x_0)$ and take y close to x in $\bar{B}_a(x_0)$. We let S_x be the shrinking map of the theorem, corresponding to the initial condition x. Then

$$\|\alpha_x - S_y\alpha_x\| = \|S_x\alpha_x - S_y\alpha_x\| \leqq |x - y|.$$

Let $C = bK$ so $0 < C < 1$. Then

$$\|\alpha_x - S_y^n\alpha_x\| \leqq \|\alpha_x - S_y\alpha_x\| + \|S_y\alpha_x - S_y^2\alpha_x\| + \cdots + \|S_y^{n-1}\alpha_x - S_y^n\alpha_x\|$$
$$\leqq (1 + C + \cdots + C^{n-1})|x - y|.$$

Since the limit of $S_y^n\alpha_x$ is equal to α_y as n goes to infinity, the continuity of the map $x \mapsto \alpha_x$ follows at once. In fact, the map satisfies a Lipschitz condition as stated.

It is easy to formulate a uniqueness theorem for integral curves over their whole domain of definition.

Theorem 1.3 (Uniqueness Theorem). *Let U be open in $\mathbf{E}$ and let $f: U \to E$ be a vector field of class C^p, $p \geq 1$. Let*

$$\alpha_1: J_1 \to U \qquad and \qquad \alpha_2: J_2 \to U$$

be two integral curves for f with the same initial condition x_0. Then α_1 and α_2 are equal on $J_1 \cap J_2$.

Proof. Let Q be the set of numbers b such that $\alpha_1(t) = \alpha_2(t)$ for

$$0 \leq t < b.$$

Then Q contains some number $b > 0$ by the local uniqueness theorem. If Q is not bounded from above, the equality of $\alpha_1(t)$ and $\alpha_2(t)$ for all $t > 0$ follows at once. If Q is bounded from above, let b be its least upper bound. We must show that b is the right end point of $J_1 \cap J_2$. Suppose that this is not the case. Define curves β_1 and β_2 near 0 by

$$\beta_1(t) = \alpha_1(b + t) \qquad and \qquad \beta_2(t) = \alpha_2(b + t).$$

Then β_1 and β_2 are integral curves of f with the initial conditions $\alpha_1(b)$ and $\alpha_2(b)$ respectively. The values $\beta_1(t)$ and $\beta_2(t)$ are equal for small negative t because b is the least upper bound of Q. By continuity it follows that $\alpha_1(b) = \alpha_2(b)$, and finally we see from the local uniqueness theorem that

$$\beta_1(t) = \beta_2(t)$$

for all t in some neighborhood of 0, whence α_1 and α_2 are equal in a neighborhood of b, contradicting the fact that b is a least upper bound of Q. We can argue the same way towards the left end points, and thus prove our statement.

For each $x \in U$, let $J(x)$ be the union of all open intervals containing 0 on which integral curves for f are defined, with initial condition equal to x. The uniqueness statement allows us to define the integral curve uniquely on all of $J(x)$.

Remark. The choice of 0 as the initial time value is made for convenience. From the uniqueness statement one obtains at once (making a time translation) the analogous statement for an integral curve defined on any open interval; in other words, if J_1, J_2 do not necessarily contain 0, and t_0 is a point in $J_1 \cap J_2$ such that $\alpha_1(t_0) = \alpha_2(t_0)$, and also we have the differential equations

$$\alpha_1'(t) = f(\alpha_1(t)) \qquad \text{and} \qquad \alpha_2'(t) = f(\alpha_2(t)),$$

then α_1 and α_2 are equal on $J_1 \cap J_2$.

In practice, one meets vector fields which may be time dependent, and also depend on parameters. We discuss these to show that their study reduces to the study of the standard case.

Time-dependent vector fields

Let J be an open interval, U open in a normed vector space $\mathbf{E}$, and

$$f: J \times U \to \mathbf{E}$$

a C^p map, which we view as depending on time $t \in J$. Thus for each t, the map $x \mapsto f(t, x)$ is a vector field on U. Define

$$\bar{f}: J \times U \to \mathbf{R} \times \mathbf{E}$$

by

$$\bar{f}(t, x) = (1, f(t, x)),$$

and view $\bar{f}$ as a time-independent vector field on $J \times U$. Let $\bar{\alpha}$ be its flow, so that

$$\bar{\alpha}'(t, s, x) = \bar{f}(\bar{\alpha}(t, s, x)), \quad \bar{\alpha}(0, s, x) = (s, x).$$

We note that $\bar{\alpha}$ has its values in $J \times U$ and thus can be expressed in terms of two components. In fact, it follows at once that we can write $\bar{\alpha}$ in the form

$$\bar{\alpha}(t, s, x) = (t + s, \bar{\alpha}_2(t, s, x)).$$

Then $\bar{\alpha}_2$ satisfies the differential equation

$$D_1\bar{\alpha}_2(t, s, x) = f(t + s, \bar{\alpha}_2(t, s, x))$$

as we see from the definition of $\bar{f}$. Let

$$\beta(t, x) = \bar{\alpha}_2(t, 0, x).$$

Then β is a flow for f, that is β satisfies the differential equation

$$D_1\beta(t, x) = f(t, \beta(t, x)), \quad \beta(0, x) = x.$$

Given $x \in U$, any value of t such that α is defined at (t, x) is also such that $\bar{\alpha}$ is defined at $(t, 0, x)$ because α_x and β_x are integral curves of the same vector field, with the same initial condition, hence are equal. Thus the study of time-dependent vector fields is reduced to the study of time-independent ones.

Dependence on parameters

Let V be open in some space $\mathbf{F}$ and let

$$g : J \times V \times U \to \mathbf{E}$$

be a map which we view as a time-dependent vector field on U, also depending on parameters in V. We define

$$G : J \times V \times U \to \mathbf{F} \times \mathbf{E}$$

by

$$G(t, z, y) = (0, g(t, z, y))$$

for $t \in J$, $z \in V$, and $y \in U$. This is now a time-dependent vector field on $V \times U$. A local flow for G depends on three variables, say $\beta(t, z, y)$, with initial condition $\beta(0, z, y) = (z, y)$. The map β has two components, and it is immediately clear that we can write

$$\beta(t, z, y) = (z, \alpha(t, z, y))$$

for some map α depending on three variables. Consequently α satisfies the differential equation

$$D_1\alpha(t, z, y) = g(t, z, \alpha(t, z, y)), \quad \alpha(0, z, y) = y,$$

which gives the flow of our original vector field g depending on the parameters $z \in V$. This procedure reduces the study of differential equations depending on parameters to those which are independent of parameters.

We shall now investigate the behavior of the flow with respect to its second argument, i.e. with respect to the points of U. We shall give two methods for this. The first depends on approximation estimates, and the second on the implicit mapping theorem in function spaces.

Let J_0 be an open subinterval of J containing 0, and let

$$\varphi : J_0 \to U$$

be of class C^1. We shall say that φ is an ϵ-**approximate solution** of f on J_0 if

$$\left|\varphi'(t) - f\big(t, \varphi(t)\big)\right| \leq \epsilon$$

for all t in J_0.

Proposition 1.4. *Let φ_1 and φ_2 be two ϵ_1- and ϵ_2-approximate solutions of f on J_0 respectively, and let $\epsilon = \epsilon_1 + \epsilon_2$. Assume that f is Lipschitz with constant K on U uniformly in J_0, or that $D_2 f$ exists and is bounded by K on $J \times U$. Let t_0 be a point of J_0. Then for any t in J_0, we have*

$$|\varphi_1(t) - \varphi_2(t)| \leq |\varphi_1(t_0) - \varphi_2(t_0)|e^{K|t-t_0|} + \frac{\epsilon}{K}e^{K|t-t_0|}.$$

Proof. By assumption, we have

$$\left|\varphi_1'(t) - f\big(t, \varphi_1(t)\big)\right| \leq \epsilon_1,$$
$$\left|\varphi_2'(t) - f\big(t, \varphi_2(t)\big)\right| \leq \epsilon_2.$$

From this we get

$$\left|\varphi_1'(t) - \varphi_2'(t) + f\big(t, \varphi_2(t)\big) - f\big(t, \varphi_1(t)\big)\right| \leq \epsilon.$$

Say $t \geq t_0$ to avoid putting bars around $t - t_0$. Let

$$\psi(t) = |\varphi_1(t) - \varphi_2(t)|,$$
$$\omega(t) = \left|f\big(t, \varphi_1(t)\big) - f\big(t, \varphi_2(t)\big)\right|.$$

Then, after integrating from t_0 to t, and using triangle inequalities we obtain

$$|\psi(t) - \psi(t_0)| \leq \epsilon(t - t_0) + \int_{t_0}^t \omega(u)\, du$$

$$\leq \epsilon(t - t_0) + K\int_{t_0}^t \psi(u)\, du$$

$$\leq K\int_{t_0}^t [\psi(u) + \epsilon/K]\, du,$$

and finally the recurrence relation

$$\psi(t) \leq \psi(t_0) + K\int_{t_0}^t [\psi(u) + \epsilon/K]\, du.$$

On any closed subinterval of J_0, our map ψ is bounded. If we add ϵ/K to both sides of this last relation, then we see that our proposition will follow from the next lemma.

Lemma 1.5. *Let g be a positive real valued function on an interval, bounded by a number L. Let t_0 be in the interval, say $t_0 \leq t$, and assume that there are numbers A, $K \geq 0$ such that*

$$g(t) \leq A + K \int_{t_0}^{t} g(u) \, du.$$

Then for all integers $n \geq 1$ we have

$$g(t) \leq A\left[1 + \frac{K(t - t_0)}{1!} + \cdots + \frac{K^{n-1}(t - t_0)^{n-1}}{(n-1)!}\right] + \frac{LK^n(t - t_0)^n}{n!}.$$

Proof. The statement is an assumption for $n = 1$. We proceed by induction. We integrate from t_0 to t, multiply by K, and use the recurrence relation. The statement with $n + 1$ then drops out of the statement with n.

Corollary 1.6. *Let $f: J \times U \to E$ be continuous, and satisfy a Lipschitz condition on U uniformly with respect to J. Let x_0 be a point of U. Then there exists an open subinterval J_0 of J containing 0, and an open subset of U containing x_0 such that f has a unique flow*

$$\alpha: J_0 \times U_0 \to U.$$

We can select J_0 and U_0 such that α is continuous and satisfies a Lipschitz condition on $J_0 \times U_0$.

Proof. Given x, y in U_0 we let $\varphi_1(t) = \alpha(t, x)$ and $\varphi_2(t) = \alpha(t, y)$, using Proposition 1.6 to get J_0 and U_0. Then $\epsilon_1 = \epsilon_2 = 0$. For s, t in J_0 we obtain

$$|\alpha(t, x) - \alpha(s, y)| \leq |\alpha(t, x) - \alpha(t, y)| + |\alpha(t, y) - \alpha(s, y)|$$
$$\leq |x - y|e^K + |t - s|L,$$

if we take J_0 of small length, and L is a bound for f. Indeed, the term containing $|x - y|$ comes from Proposition 1.4, and the term containing $|t - s|$ comes from the definition of the integral curve by means of an integral and the bound L for f. This proves our corollary.

Corollary 1.7. *Let J be an open interval of $\mathbf{R}$ containing 0 and let U be open in $\mathbf{E}$. Let $f: J \times U \to \mathbf{E}$ be a continuous map, which is Lipschitz* on U uniformly for every compact subinterval of J. Let $t_0 \in J$ and let φ_1, φ_2 be two morphisms of class C^1 such that $\varphi_1(t_0) = \varphi_2(t_0)$ and satisfying the relation

$$\varphi'(t) = f\big(t, \varphi(t)\big)$$

for all t in J. Then $\varphi_1(t) = \varphi_2(t)$.

Proof. We can take $\epsilon = 0$ in the proposition.

The above corollary gives us another proof for the uniqueness of integral curves. Given $f \colon J \times U \to E$ as in this corollary, we can define an integral curve α for f on a maximal open subinterval of J having a given value $\alpha(t_0)$ for a fixed t_0 in J. Let J be the open interval (a, b) and let (a_0, b_0) be the interval on which α is defined. We want to know when $b_0 = b$ (or $a_0 = a$), that is when the integral curve of f can be continued to the entire interval over which f itself is defined.

There are essentially two reasons why it is possible that the integral curve cannot be extended to the whole domain of definition J, or cannot be extended to infinity in case f is independent of time. One possibility is that the integral curve tends to get out of the open set U, as on the following picture:

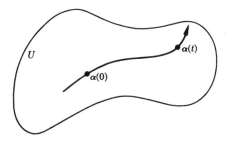

This means that as t approaches b_0, say, the curve $\alpha(t)$ approaches a point which does not lie in U. Such an example can actually be constructed artificially. If we are in a situation when a curve can be extended to infinity, just remove a point from the open set lying on the curve. Then the integral curve on the resulting open set cannot be continued to infinity. The second possibility is that the vector field is unbounded. The next corollary shows that these possibilities are the only ones. In other words, if an integral curve does not tend to get out of the open set, and if the vector field is bounded, then the curve can be continued as far as the original data will allow a priori.

Corollary 1.8. *Let J be the open interval (a, b) and let U be open in $\mathbf{E}$. Let $f \colon J \times U \to \mathbf{E}$ be a continuous map, which is Lipschitz on U,* *uniformly for every compact subset of J. Let α be an integral curve of f, defined on a maximal open subinterval (a_0, b_0) of J. Assume:*

(i)   *There exists $\epsilon > 0$ such that $\overline{\alpha((b_0 - \epsilon, b_0))}$ is contained in U.*
(ii) *There exists a number $B > 0$ such that $|f(t, \alpha(t))| \leqq B$ for all t in $(b_0 - \epsilon, b_0)$.*

Then $b_0 = b$.

Proof. From the integral expression for α, namely

$$\alpha(t) = \alpha(t_0) + \int_{t_0}^{t} f(u, \alpha(u))\, du,$$

we see that for t_1, t_2 in $(b_0 - \epsilon, b_0)$ we have

$$|\alpha(t_1) - \alpha(t_2)| \leqq B|t_1 - t_2|.$$

From this it follows that the limit

$$\lim_{t \to b_0} \alpha(t)$$

exists, and is equal to an element x_0 of U (by hypothesis (i)). Assume that $b_0 \neq b$. By the local existence theorem, there exists an integral curve β of f defined on an open interval containing b_0 such that $\beta(b_0) = x_0$ and $\beta'(t) = f(t, \beta(t))$. Then $\beta' = \alpha'$ on an open interval to the left of b_0, and hence α, β differ by a constant on this interval. Since their limit as $t \to b_0$ are equal, this constant is 0. Thus we have extended the domain of definition of α to a larger interval, as was to be shown.

The next proposition describes the solutions of **linear differential equations** depending on parameters.

Proposition 1.9. *Let J be an open interval of* **R** *containing 0, and let V be an open set in a vector space. Let* **E** *be a vector space. Let*

$$g: J \times V \to L(\mathbf{E}, \mathbf{E})$$

be a continuous map. Then there exists a unique map

$$\lambda: J \times V \to L(\mathbf{E}, \mathbf{E})$$

which, for each $x \in V$, is a solution of the differential equation

$$D_1\lambda(t, x) = g(t, x)\lambda(t, x), \quad \lambda(0, x) = \mathrm{id}.$$

This map λ is continuous.

Remark. In the present case of a linear differential equation, it is not necessary to shrink the domain of definition of its flow. Note that the differential equation is on the space of linear maps. The corresponding linear equation on **E** itself will come out as a corollary.

Proof of Proposition 1.9. Let us first fix $x \in V$. Consider the differential equation

$$D_1\lambda(t, x) = g(t, x)\lambda(t, x),$$

with initial condition $\lambda(0, x) = \text{id}$. This is a differential equation on $L(\mathbf{E}, \mathbf{E})$, where $f(t, z) = g_x(t)z$ for $z \in L(\mathbf{E}, \mathbf{E})$, and we write $g_x(t)$ instead of $g(t, x)$. Let the notation be as in Corollary 1.8. Then hypothesis (i) is automatically satisfied since the open set U is all of $L(\mathbf{E}, \mathbf{E})$. On every compact subinterval of J, g_x is bounded, being continuous. Omitting the index x for simplicity, we have

$$\lambda(t) = \text{id} + \int_0^t g(u)\lambda(u)\, du,$$

whence for $t \geq 0$, say

$$|\lambda(t)| \leq 1 + B \int_0^t |\lambda(u)|\, du.$$

Using Lemma 1.5, we see that hypothesis (ii) of Corollary 1.8 is also satisfied. Hence the integral curve is defined on all of J.

We shall now prove the continuity of λ. Let $(t_0, x_0) \in J \times V$. Let I be a compact interval contained in J, and containing t_0 and 0. As a function of t, $\lambda(t, x_0)$ is continuous (even differentiable). Let $C > 0$ be such that $|\lambda(t, x_0)| \leq C$ for all $t \in I$. Let V_1 be an open neighborhood of x_0 in V such that g is bounded by a constant $K > 0$ on $I \times V_1$.

For $(t, x) \in I \times V_1$ we have

$$|\lambda(t, x) - \lambda(t_0, x_0)| \leq |\lambda(t, x) - \lambda(t, x_0)| + |\lambda(t, x_0) - \lambda(t_0, x_0)|.$$

The second term on the right is small when t is close to t_0. We investigate the first term on the right, and shall estimate it by viewing $\lambda(t, x)$ and $\lambda(t, x_0)$ as approximate solutions of the differential equation satisfied by $\lambda(t, x)$. We find

$$|D_1\lambda(t, x_0) - g(t, x)\lambda(t, x_0)|$$

$$= |D_1\lambda(t, x_0) - g(t, x)\lambda(t, x_0) + g(t, x_0)\lambda(t, x_0) - g(t, x_0)\lambda(t, x_0)|$$

$$\leq |g(t, x_0) - g(t, x)|\,|\lambda(t, x_0)| \leq |g(t, x_0) - g(t, x)|C.$$

By the usual proof of uniform continuity applied to the compact set $I \times \{x_0\}$, given $\epsilon > 0$, there exists an open neighborhood V_0 of x_0 contained in V_1, such that for all $(t, x) \in I \times V_0$ we have

$$|g(t, x) - g(t, x_0)| < \epsilon/C.$$

This implies that $\lambda(t, x_0)$ is an ϵ-approximate solution of the differential equation satisfied by $\lambda(t, x)$. We apply Proposition 1.4 to the two curves

$$\varphi_0(t) = \lambda(t, x_0) \quad \text{and} \quad \varphi_x(t) = \lambda(t, x)$$

for each $x \in V_0$. We use the fact that $\lambda(0, x) = \lambda(0, x_0) = \text{id}$. We then find

$$|\lambda(t, x) - \lambda(t, x_0)| < \epsilon K_1$$

for some constant $K_1 > 0$, thereby proving the continuity of λ at (t_0, x_0).

Corollary 1.10. *Let the notation be as in Proposition* 1.9. *For each* $x \in V$ *and* $z \in E$ *the curve*

$$\beta(t, x, z) = \lambda(t, x)z$$

with initial condition $\beta(0, x, z) = z$ *is a solution of the differential equation*

$$D_1\beta(t, x, z) = g(t, x)\beta(t, x, z).$$

Furthermore, β *is continuous in its three variables.*

Proof. Obvious.

Theorem 1.11 (Local Smoothness Theorem). *Let* J *be an open interval in* **R** *containing* 0 *and* U *open in the vector space* **E**. *Let*

$$f \colon J \times U \to \mathbf{E}$$

be a C^p-*morphism with* $p \geq 1$, *and let* $x_0 \in U$. *There exists a unique local flow for* f *at* x_0. *We can select an open subinterval* J_0 *of* J *containing* 0 *and an open subset* U_0 *of* U *containing* x_0 *such that the unique local flow*

$$\alpha \colon J_0 \times U_0 \to U$$

is of class C^p, *and such that* $D_2\alpha$ *satisfies the differential equation*

$$\boxed{D_1 D_2\alpha(t, x) = D_2 f\big(t, \alpha(t, x)\big) D_2\alpha(t, x)}$$

on $J_0 \times U_0$ *with initial condition* $D_2\alpha(0, x) = \text{id}.$

Proof. Let

$$g: J \times U \to L(\mathbf{E}, \mathbf{E})$$

be given by $g(t, x) = D_2 f(t, \alpha(t, x))$. Select J_1 and U_0 such that α is bounded and Lipschitz on $J_1 \times U_0$ (by Corollary 1.6), and such that g is continuous and bounded on $J_1 \times U_0$. Let J_0 be an open subinterval of J_1 containing 0 such that its closure $\bar{J}_0$ is contained in J_1.

Let $\lambda(t, x)$ be the solution of the differential equation on $L(\mathbf{E}, \mathbf{E})$ given by

$$D_1 \lambda(t, x) = g(t, x)\lambda(t, x), \qquad \lambda(0, x) = \mathrm{id},$$

as in Proposition 1.9. We contend that $D_2 \alpha$ exists and is equal to λ on $J_0 \times U_0$. This will prove that $D_2 \alpha$ is continuous, on $J_0 \times U_0$.

Fix $x \in U_0$. Let

$$\theta(t, h) = \alpha(t, x + h) - \alpha(t, x).$$

Then

$$\begin{aligned}
D_1 \theta(t, h) &= D_1 \alpha(t, x + h) - D_1 \alpha(t, x) \\
&= f(t, \alpha(t, x + h)) - f(t, \alpha(t, x)).
\end{aligned}$$

By the mean value theorem, we obtain

$$\begin{aligned}
&|D_1 \theta(t, h) - g(t, x)\theta(t, h)| \\
&= |f(t, \alpha(t, x + h)) - f(t, \alpha(t, x)) - D_2 f(t, \alpha(t, x))\theta(t, h)| \\
&\leq |h| \sup |D_2 f(t, y) - D_2 f(t, \alpha(t, x))|,
\end{aligned}$$

where y ranges over the segment between $\alpha(t, x)$ and $\alpha(t, x + h)$. By the compactness of $\bar{J}_0$ it follows that our last expression is bounded by $|h|\psi(h)$ where $\psi(h)$ tends to 0 with h, uniformly for t in $\bar{J}_0$. Hence we obtain

$$|\theta'(t, h) - g(t, x)\theta(t, h)| \leq |h|\psi(h),$$

for all t in $\bar{J}_0$. This shows that $\theta(t, h)$ is an $|h|\psi(h)$ approximate solution for the differential equation satisfied by $\lambda(t, x)h$, namely

$$D_1 \lambda(t, x)h - g(t, x)\lambda(t, x)h = 0,$$

with the initial condition $\lambda(0, x)h = h$. We note that $\theta(t, h)$ has the same initial condition, $\theta(0, h) = h$. Taking $t_0 = 0$ in Proposition 1.4, we obtain the estimate

$$|\theta(t, h) - \lambda(t, x)h| \leq C_1 |h|\psi(h)$$

for all t in $\bar{J}_0$. This proves that $D_2\alpha$ is equal to λ on $J_0 \times U_0$, and is therefore continuous on $J_0 \times U_0$.

We have now proved that $D_1\alpha$ and $D_2\alpha$ exist and are continuous on $J_0 \times U_0$, and hence that α is of class C^1 on $J_0 \times U_0$.

Furthermore, $D_2\alpha$ satisfies the differential equation given in the statement of our theorem on $J_0 \times U_0$. Thus our theorem is proved when $p = 1$.

A flow which satisfies the properties stated in the theorem will be called **locally of class** C^p.

Consider now again the linear equation of Proposition 1.9. We reformulate it to eliminate formally the parameters, namely we define a vector field

$$G: \ J \times V \times L(\mathbf{E}, \mathbf{E}) \to F \times L(\mathbf{E}, \mathbf{E})$$

to be the map such that

$$G(t, \ x, \ \omega) = \big(0, \ g(t, \ x)\omega\big)$$

for $\omega \in L(\mathbf{E}, \mathbf{E})$. The flow for this vector field is then given by the map Λ such that

$$\Lambda(t, \ x, \ \omega) = \big(x, \ \lambda(t, \ x)\omega\big).$$

If g is of class C^1 we can now conclude that the flow Λ is locally of class C^1, and hence putting $\omega = \mathrm{id}$, that λ is locally of class C^1.

We apply this to the case when $g(t, x) = D_2 f\big(t, \ \alpha(t, \ x)\big)$, and to the solution $D_2\alpha$ of the differential equation

$$D_1(D_2\alpha)(t, \ x) = g(t, \ x)D_2\alpha(t, \ x)$$

locally at each point $(0, \ x)$, $x \in U$. Let $p \geq 2$ be an integer and assume out theorem proved up to $p - 1$, so that we can assume α locally of class C^{p-1}, and f of class C^p. Then g is locally of class C^{p-1}, whence $D_2\alpha$ is locally C^{p-1}. From the expression

$$D_1\alpha(t, \ x) = f\big(t, \ \alpha(t, \ x)\big)$$

we conclude that $D_1\alpha$ is C^{p-1}, whence α is locally C^p.

If f is C^∞, and if we knew that α is of class C^p for every integer p **on its domain of definition**, then we could conclude that α is C^∞; in other words, there is no shrinkage in the inductive application of the local theorem.

We now give the arguments needed to globalize the smoothness. We may limit ourselves to the time-independent case. We have seen that the time-dependent case reduces to the other.

Let U be open in a vector space $\mathbf{E}$, and let $f: U \to \mathbf{E}$ be a C^p vector field. We let $J(x)$ be the domain of the integral curve with initial condition equal to w.

Let $\mathfrak{D}(f)$ be the set of all points (t, x) in $\mathbf{R} \times U$ such that t lies in $J(x)$. Then we have a map

$$\alpha: \mathfrak{D}(f) \to U$$

defined on all of $\mathfrak{D}(f)$, letting $\alpha(t, x) = \alpha_x(t)$ be the integral curve on $J(x)$ having x as initial condition. We call this the **flow** determined by f, and we call $\mathfrak{D}(f)$ its **domain of definition**.

Lemma 1.12. *Let $f: U \to E$ be a C^p vector field on the open set U of E, and let α be its flow. Abbreviate $\alpha(t, x)$ by tx, if (t, x) is in the domain of definition of the flow. Let $x \in U$. If t_0 lies in $J(x)$, then*

$$J(t_0 x) = J(x) - t_0$$

(translation of $J(x)$ by $-t_0$), and we have for all t in $J(x) - t_0$:

$$t(t_0 x) = (t + t_0)x.$$

Proof. The two curves defined by

$$t \mapsto \alpha\big(t, \alpha(t_0, x)\big) \qquad \text{and} \qquad t \mapsto \alpha(t + t_0, x)$$

are integral curves of the same vector field, with the same initial condition $t_0 x$ at $t = 0$. Hence they have the same domain of definition $J(t_0 x)$. Hence t_1 lies in $J(t_0 x)$ if and only if $t_1 + t_0$ lies in $J(x)$. This proves the first assertion. The second assertion comes from the uniqueness of the integral curve having given initial condition, whence the theorem follows.

Theorem 1.13 (Global Smoothness of the Flow). *If f is of class C^p (with $p \leq \infty$), then its flow is of class C^p on its domain of definition.*

Proof. First let p be an integer ≥ 1. We know that the flow is locally of class C^p at each point $(0, x)$, by the local theorem. Let $x_0 \in U$ and let $J(x_0)$ be the maximal interval of definition of the integral curve having x_0 as initial condition. Let $\mathfrak{D}(f)$ be the domain of definition of the flow, and let α be the flow. Let Q be the set of numbers $b > 0$ such that for each t with $0 \leq t < b$ there exists an open interval J containing t and an open set V containing x_0 such that $J \times V$ is contained in $\mathfrak{D}(f)$ and such that α is of class C^p on $J \times V$. Then Q is not empty by the local theorem. If Q is not bounded from above, then we are done looking toward the right end point of $J(x_0)$. If Q is bounded from above, we let b be its least upper bound. We must prove that b is the right end point of $J(x_0)$. Suppose that this is not the case. Then $\alpha(b, x_0)$ is defined. Let $x_1 = \alpha(b, x_0)$. By the local theorem, we have a unique local flow at x_1, which we denote by β:

$$\beta: \ J_a \times B_a(x_1) \to U, \qquad \beta(0, x) = x,$$

defined for some open interval $J_a = (-a, a)$ and open ball $B_a(x_1)$ of radius a centered at x_1. Let δ be so small that whenever $b - \delta < t < b$ we have

$$\alpha(t, x_0) \in B_{a/4}(x_1).$$

We can find such δ because

$$\lim_{t \to b} \alpha(t, x_0) = x_1$$

by continuity. Select a point t_1 such that $b - \delta < t_1 < b$. By the hypothesis on b, we can select an open interval J_1 containing t_1 and an open set U_1 containing x_0 so that

$$\alpha: \ J_1 \times U_1 \to B_{a/2}(x_1)$$

maps $J_1 \times U_1$ into $B_{a/2}(x_1)$. We can do this because α is continuous at (t_1, x_0), being in fact C^p at this point. If $|t - t_1| < a$ and $x \in U_1$, we define

$$\varphi(t, x) = \beta\big(t - t_1, \alpha(t_1, x)\big).$$

Then

$$\varphi(t_1, x) = \beta\big(0, \alpha(t_1, x)\big) = \alpha(t_1, x)$$

and

$$\begin{aligned}
D_1\varphi(t, x) &= D_1\beta\big(t - t_1, \alpha(t_1, x)\big) \\
&= f\big(\beta(t - t_1, \alpha(t_1, x))\big) \\
&= f\big(\varphi(t, x)\big).
\end{aligned}$$

Hence both φ_x and α_x are integral curves for f with the same value at t_1. They coincide on any interval on which they are defined by the uniqueness theorem. If we take δ very small compared to a, say $\delta < a/4$, we see that φ is an extension of α to an open set containing (t_1, α_0), and also containing (b, x_0). Furthermore, φ is of class C^p, thus contradicting the fact that b is strictly smaller than the end point of $J(x_0)$. Similarly, one proves the analogous statement on the other side, and we therefore see that $\mathfrak{D}(f)$ is open in $\mathbf{R} \times U$ and that α is of class C^p on $\mathfrak{D}(f)$, as was to be shown.

The idea of the above proof is very simple geometrically. We go as far to the right as possible in such a way that the given flow α is of class C^p locally at (t, x_0). At the point $\alpha(b, x_0)$ we then use the flow β to extend differentiably the flow α in case b is not the right-hand point of $J(x_0)$. The flow β at $\alpha(b, x_0)$ has a fixed local domain of definition, and we simply take t close enough to b so that β gives an extension of α, as described in the above proof.

Of course, if f is of class C^∞, then we have shown that α is of class C^p for each positive integer p, and therefore the flow is also of class C^∞.

In the next section, we shall see how these arguments globalize even more to manifolds.

IV, §2. VECTOR FIELDS, CURVES, AND FLOWS

Let X be a manifold of class C^p with $p \geq 2$. Let $\pi: T(X) \to X$ be its tangent bundle. Then $T(X)$ is of class C^{p-1}, $p - 1 \geq 1$.

By a (time-independent) **vector field** on X we mean a cross section of the tangent bundle, i.e. a morphism (of class C^{p-1})

$$\xi: X \to T(X)$$

such that $\xi(x)$ lies in the tangent space $T_x(X)$ for each $x \in X$, or in other words, such that $\pi\xi = \text{id}$.

If $T(X)$ is trivial, and say X is an **E**-manifold, so that we have a VB-isomorphism of $T(X)$ with $X \times \mathbf{E}$, then the morphism ξ is completely determined by its projection on the second factor, and we are essentially in the situation of the preceding paragraph, except for the fact that our vector field is independent of time. In such a product representation, the projection of ξ on the second factor will be called the **local representation** of ξ. It is a C^{p-1}-morphism

$$f: X \to \mathbf{E}$$

and $\xi(x) = \big(x, f(x)\big)$. We shall also say that ξ is **represented by f locally** if we work over an open subset U of X over which the tangent bundle admits a trivialisation. We then frequently use ξ itself to denote this local representation.

Let J be an open interval of $\mathbf{R}$. The tangent bundle of J is then $J \times \mathbf{R}$ and we have a canonical section ι such that $\iota(t) = 1$ for all $t \in J$. We sometimes write ι_t instead of $\iota(t)$.

By a **curve** in X we mean a morphism (always of class $\geqq 1$ unless otherwise specified)

$$\alpha\colon J \to X$$

from an open interval in $\mathbf{R}$ into X. If $g\colon X \to Y$ is a morphism, then $g \circ \alpha$ is a curve in Y. From a given curve α, we get an induced map on the tangent bundles:

$$
\begin{array}{ccc}
J \times \mathbf{R} & \xrightarrow{\ \alpha_* \ } & T(X) \\
\downarrow & & \downarrow{\scriptstyle \pi} \\
J & \xrightarrow[\ \alpha \ ]{} & X
\end{array}
$$

and $\alpha_* \circ \iota$ will be denoted by α' or by $d\alpha/dt$ if we take its value at a point t in J. Thus α' is a curve in $T(X)$, of class C^{p-1} if α is of class C^p. Unless otherwise specified, it is always understood in the sequel that we start with enough differentiability to begin with so that we never end up with maps of class < 1. Thus to be able to take derivatives freely we have to take X and α of class C^p with $p \geqq 2$.

If $g\colon X \to Y$ is a morphism, then

$$(g \circ \alpha)'(t) = g_* \alpha'(t).$$

This follows at once from the functoriality of the tangent bundle and the definitions.

Suppose that J contains 0, and let us consider curves defined on J and such that $\alpha(0)$ is equal to a fixed point x_0. We could say that two such curves α_1, α_2 are **tangent** at 0 if $\alpha_1'(0) = \alpha_2'(0)$. The reader will verify immediately that there is a natural bijection between tangency classes of curves with $\alpha(0) = x_0$ and the tangent space $T_{x_0}(X)$ of X at x_0. The tangent space could therefore have been defined alternatively by taking equivalence classes of curves through the point.

Let ξ be a vector field on X and x_0 a point of X. An **integral curve** for the vector field ξ with **initial condition** x_0, or starting at x_0, is a curve (of class C^{p-1})

$$\alpha\colon J \to X$$

mapping an open interval J of $\mathbf{R}$ containing 0 into X, such that $\alpha(0) = x_0$ and such that

$$\alpha'(t) = \xi\big(\alpha(t)\big)$$

for all $t \in J$. Using a local representation of the vector field, we know from the preceding section that integral curves exist locally. The next theorem gives us their global existence and uniqueness.

Theorem 2.1. *Let* $\alpha_1: J_1 \to X$ *and* $\alpha_2: J_2 \to X$ *be two integral curves of the vector field* ξ *on* X, *with the same initial condition* x_0. *Then* α_1 *and* α_2 *are equal on* $J_1 \cap J_2$.

Proof. Let J^* be the set of points t such that $\alpha_1(t) = \alpha_2(t)$. Then J^* certainly contains a neighborhood of 0 by the local uniqueness theorem. Furthermore, since X is Hausdorff, we see that J^* is closed. We must show that it is open. Let t^* be in J^* and define β_1, β_2 near 0 by

$$\beta_1(t) = \alpha_1(t^* + t),$$
$$\beta_2(t) = \alpha_2(t^* + t).$$

Then β_1 and β_2 are integral curves of ξ with initial condition $\alpha_1(t^*)$ and $\alpha_2(t^*)$ respectively, so by the local uniqueness theorem, β_1 and β_2 agree in a neighborhood of 0 and thus α_1, α_2 agree in a neighborhood of t^*, thereby proving our theorem.

It follows from Theorem 2.1 that the union of the domains of all integral curves of ξ with a given initial condition x_0 is an open interval which we denote by $J(x_0)$. Its end points are denoted by $t^+(x_0)$ and $t^-(x_0)$ respectively. (We do not exclude $+\infty$ and $-\infty$.)

Let $\mathfrak{D}(\xi)$ be the subset of $\mathbf{R} \times X$ consisting of all points (t, x) such that

$$t^-(x) < t < t^+(x).$$

A (global) **flow** for ξ is a mapping

$$\alpha: \mathfrak{D}(\xi) \to X,$$

such that for each $x \in X$, the map $\alpha_x: J(x) \to X$ given by

$$\alpha_x(t) = \alpha(t, x)$$

defined on the open interval $J(x)$ is a morphism and is an integral curve for ξ with initial condition x. When we select a chart at a point x_0 of X, then one sees at once that this definition of flow coincides with the definition we gave locally in the previous section, for the local representation of our vector field.

Given a point $x \in X$ and a number t, we say that tx is **defined** if (t, x) is in the domain of α, and we denote $\alpha(t, x)$ by tx in that case.

Theorem 2.2. *Let* ξ *be a vector field on* X, *and* α *its flows. Let* x *be a point of* X. *If* t_0 *lies in* $J(x)$, *then*

$$J(t_0 x) = J(x) - t_0$$

(*translation of* $J(x)$ *by* $-t_0$), *and we have for all* t *in* $J(x) - t_0$:

$$t(t_0 x) = (t + t_0)x.$$

Proof. Our first assertion follows immediately from the maximality assumption concerning the domains of the integral curves. The second is equivalent to saying that the two curves given by the left-hand side and right-hand side of the last equality are equal. They are both integral curves for the vector field, with initial condition $t_0 x$ and must therefore be equal.

In particular, if t_1, t_2 are two numbers such that $t_1 x$ is defined and $t_2(t_1 x)$ is also defined, then so is $(t_1 + t_2)x$ and they are equal.

Theorem 2.3. *Let* ξ *be a vector field on* X, *and* x *a point of* X. *Assume that* $t^+(x) < \infty$. *Given a compact set* $A \subset X$, *there exists* $\epsilon > 0$ *such that for all* $t > t^+(x) - \epsilon$, *the point* tx *does not lie in* A, *and similarly for* t^-.

Proof. Suppose such ϵ does not exist. Then we can find a sequence t_n of real numbers approaching $t^+(x)$ from below, such that $t_n x$ lies in A. Since A is compact, taking a subsequence if necessary, we may assume that $t_n x$ converges to a point in A. By the local existence theorem, there exists a neighborhood U of this point y and a number $\delta > 0$ such that $t^+(z) > \delta$ for all $z \in U$. Taking n large, we have

$$t^+(x) < \delta + t_n$$

and $t_n x$ is in U. Then by Theorem 2.2,

$$t^+(x) = t^+(t_n x) + t_n > \delta + t_n > t^+(x)$$

contradiction.

Corollary 2.4. *If* X *is compact, and* ξ *is a vector field on* X, *then*

$$\mathfrak{D}(\xi) = \mathbf{R} \times X.$$

It is also useful to give one other criterion when $\mathfrak{D}(\xi) = \mathbf{R} \times X$, even when X is not compact. Such a criterion must involve some structure stronger than the differentiable structure (essentially a metric of some sort), because we can always dig holes in a compact manifold by taking away a point.

Proposition 2.5. *Let* $\mathbf{E}$ *be a normed vector space, and* X *an* $\mathbf{E}$-*manifold. Let* ξ *be a vector field on* X. *Assume that there exist numbers* $a > 0$ *and* $K > 0$ *such that every point* x *of* X *admits a chart* (U, φ) *at* x *such that* the local representation f of the vector field on this chart is bounded by K, and so is its derivative f'. Assume also that φU contains a ball of radius a around φx. Then $\mathfrak{D}(\xi) = \mathbf{R} \times X$.

Proof. This follows at once from the global continuation theorem, and the uniformity of Proposition 1.1.

We shall prove finally that $\mathfrak{D}(\xi)$ is open and that α is a morphism.

Theorem 2.6. *Let ξ be a vector field of class C^{p-1} on the C^p-manifold X $(2 \leq p \leq \infty)$. Then $\mathfrak{D}(\xi)$ is open in $\mathbf{R} \times X$, and the flow α for ξ is a C^{p-1}-morphism.*

Proof. Let first p be an integer ≥ 2. Let $x_0 \in X$. Let J^* be the set of points in $J(x_0)$ for which there exists a number $b > 0$ and an open neighborhood U of x_0 such that $(t - b, t + b) U$ is contained in $\mathfrak{D}(\xi)$, and such that the restriction of the flow α to this product is a C^{p-1}-morphism. Then J^* is open in $J(x_0)$, and certainly contains 0 by the local theorem. We must therefore show that J^* is closed in $J(x_0)$.

Let s be in its closure. By the local theorem, we can select a neighborhood V of $sx_0 = \alpha(s, x_0)$ so that we have a unique local flow

$$\beta: J_a \times V \to X$$

for some number $a > 0$, with initial condition $\beta(0, x) = x$ for all $x \in V$, and such that this local flow β is C^{p-1}.

The integral curve with initial condition x_0 is certainly continuous on $J(x_0)$. Thus tx_0 approaches sx_0 as t approaches s. Let V_1 be a given small neighborhood of sx_0 contained in V. By the definition of J^*, we can find an element t_1 in J^* very close to s, and a small number b (compared to a) and a small neighborhood U of x_0 such that α maps the product

$$(t_1 - b, t_1 + b) \times U$$

into V_1, and is C^{p-1} on this product. For $t \in J_a + t_1$ and $x \in U$, we define

$$\varphi(t, x) = \beta\big(t - t_1, \alpha(t_1, x)\big).$$

Then $\varphi(t_1, x) = \beta\big(0, \alpha(t_1, x)\big) = \alpha(t_1, x)$, and

$$D_1\varphi(t, x) D_1\beta(t - t_1, \alpha(t_1, x))$$
$$= \xi\big(\beta(t - t_1, \alpha(t_1, x))\big)$$
$$= \xi\big(\varphi(t, x)\big).$$

Hence both φ_x, α_x are integral curves for ξ, with the same value at t_1. They coincide on any interval on which they are defined, so that φ_x is a continuation of α_x to a bigger interval containing s. Since α is C^{p-1} on the product $(t_1 - b, t_1 + b) \times U$, we conclude that φ is also C^{p-1} on $(J_a + t_1) \times U$. From this we see that $\mathfrak{D}(\xi)$ is open in $\mathbf{R} \times X$, and that α is of class C^{p-1} on its full domain $\mathfrak{D}(\xi)$. If $p = \infty$, then we can now conclude that α is of class C^r for each positive integer r on $\mathfrak{D}(\xi)$, and hence is C^∞, as desired.

Corollary 2.7. *For each $t \in \mathbf{R}$, the set of $x \in X$ such that (t, x) is contained in the domain $\mathfrak{D}(\xi)$ is open in X.*

Corollary 2.8. *The functions $t^+(x)$ and $t^-(x)$ are upper and lower semicontinuous respectively.*

Theorem 2.9. *Let ξ be a vector field on X and α its flow. Let $\mathfrak{D}_t(\xi)$ be the set of points x of X such that (t, x) lies in $\mathfrak{D}(\xi)$. Then $\mathfrak{D}_t(\xi)$ is open for each $t \in \mathbf{R}$, and α_t is an isomorphism of $\mathfrak{D}_t(\xi)$ onto an open subset of X. In fact, $\alpha_t(\mathfrak{D}_t) = \mathfrak{D}_{-t}$ and $\alpha_t^{-1} = \alpha_{-t}$.*

Proof. Immediate from the preceding theorem.

Corollary 2.10. *If x_0 is a point of X and t is in $J(x_0)$, then there exists an open neighborhood U of x_0 such that t lies in $J(x)$ for all $x \in U$, and the map*

$$x \mapsto tx$$

is an isomorphism of U onto an open neighborhood of tx_0.

Critical points

Let ξ be a vector field. A **critical point** of ξ is a point x_0 such that $\xi(x_0) = 0$. Critical points play a significant role in the study of vector fields, notably in the Morse theory. We don't go into this here, but just make a few remarks to show at the basic level how they affect the behavior of integral curves.

Proposition 2.11. *If α is an integral curve of a C^1 vector field, ξ, and α passes through a critical point, then α is constant, that is $\alpha(t) = x_0$ for all t.*

Proof. The constant curve through x_0 is an integral curve for the vector field, and the uniqueness theorem shows that it is the only one.

Some smoothness of the vector field in addition to continuity must be assumed for the uniqueness. For instance, the following picture illustrates a situation where the integral curves are not unique. They consist in translations of the curve $y = x^3$ in the plane. The vector field is continuous but not locally Lipschitz.

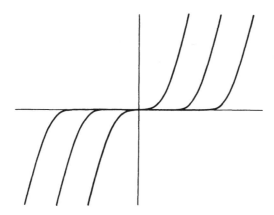

Proposition 2.12. *Let ξ be a vector field and α an integral curve for ξ. Assume that all $t \geq 0$ are in the domain of α, and that*

$$\lim_{t \to 0} \alpha(t) = x_1$$

exists. Then x_1 is a critical point for ξ, that is $\xi(x_1) = 0$.

Proof. Selecting t large, we may assume that we are dealing with the local representation f of the vector field near x_1. Then for $t' > t$ large, we have

$$\alpha(t') - \alpha(t) = \int_t^{t'} f(\alpha(u)) \, du.$$

Write $f(\alpha(u)) = f(x_1) + g(u)$, where $\lim g(u) = 0$. Then

$$|f(x_1)| \, |t' - t| \leq |\alpha(t') - \alpha(t)| + |t' - t| \sup|g(u)|,$$

where the sup is taken for u large, and hence for small values of $g(u)$. Dividing by $|t' - t|$ shows that $f(x_1)$ is arbitrarily small, hence equal to 0, as was to be shown.

Proposition 2.13. *Suppose on the other hand that x_0 is not a critical point of the vector field ξ. Then there exists a chart at x_0 such that the local representation of the vector field on this chart is constant.*

Proof. In an arbitrary chart the vector field has a representation as a morphism

$$\xi \colon U \to E$$

near x_0. Let α be its flow. We wish to "straighten out" the integral curves of the vector field according to the next figure.

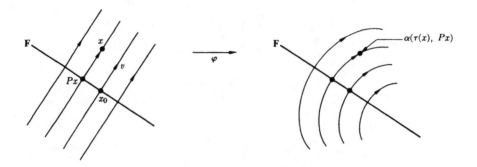

In other words, let $v = \xi(x_0)$. We want to find a local isomorphism φ at x_0 such that

$$\varphi'(x)v = \xi\big(\varphi(x)\big).$$

We inspire ourselves from the picture. Without loss of generality, we may assume that $x_0 = 0$. Let λ be a functional such that $\lambda(v) \neq 0$. We decompose E as a direct sum

$$\mathbf{E} = \mathbf{F} \oplus \mathbf{R}v,$$

where $\mathbf{F}$ is the kernel of λ. Let P be the projection on $\mathbf{F}$. We can write any x near 0 in the form

$$x = Px + \tau(x)v,$$

where

$$\tau(x) = \frac{\lambda(x)}{\lambda(v)}.$$

We then bend the picture on the left to give the picture on the right using the flow α of ξ, namely we define

$$\varphi(x) = \alpha\big(\tau(x), Px\big).$$

This means that starting at Px, instead of going linearly in the direction of v for a time $\tau(x)$, we follow the flow (integral curve) for this amount of time. We find that

$$\varphi'(x) = D_1\alpha\big(\tau(x),\, Px\big)\frac{\lambda}{\lambda(v)} + D_2\alpha\big(\tau(x),\, Px\big)\, P.$$

Hence $\varphi'(0) = \mathrm{id}$, so by the inverse mapping theorem, φ is a local isomorphism at 0. Furthermore, since $Pv = 0$ by definition, we have

$$\varphi'(x)v = D_1\alpha\big(\tau(x),\, Px\big) = \xi\big(\varphi(x)\big),$$

thus proving Proposition 2.13.

IV, §3. SPRAYS

Second-order vector fields and differential equations

Let X be a manifold of class C^p with $p \geq 3$. Then its tangent bundle $T(X)$ is of class C^{p-1}, and the tangent bundle of the tangent bundle $T\big(T(x)\big)$ is of class C^{p-2}, with $p - 2 \geq 1$.

Let $\alpha\colon J \to X$ be a curve of class C^q ($q \leq p$). A **lifting** of α into $T(X)$ is a curve $\beta\colon J \to T(X)$ such that $\pi\beta = \alpha$. We shall always deal with $q \geq 2$ so that a lift will be assumed of class $q - 1 \geq 1$. Such lifts always exist, for instance the curve α' discussed in the previous section, called the **canonical lifting** of α.

A **second-order** vector field over X is a vector field F on the tangent bundle $T(X)$ (of class C^{p-1}) such that, if $\pi\colon TX \to X$ denotes the canonical projection of $T(X)$ on X, then

$$\pi_* \circ F = \mathrm{id.}, \qquad \text{that is} \quad \pi_* F(v) = v \quad \text{for all } v \text{ in } T(X).$$

Observe that the succession of symbols makes sense, because

$$\pi_*\colon TT(X) \to T(X)$$

maps the double tangent bundle into $T(X)$ itself.

A vector field F on TX is a second-order vector field on X if and only if it satisfies the following condition: Each integral curve β of F is equal to the canonical lifting of $\pi\beta$, in other words

$$(\pi\beta)' = \beta.$$

Here, $\pi\beta$ is the canonical projection of β on X, and if we put the argument t, then our formula reads

$$(\pi\beta)'(t) = \beta(t)$$

for all t in the domain of β. The proof is immediate from the definitions, because

$$(\pi\beta)' = \pi_*\beta' = \pi_* \circ F \circ \beta$$

We then use the fact that given a vector $v \in TX$, there is an integral curve $\beta = \beta_v$ with $\beta_v(0) = v$ (initial condition v).

Let $\alpha: J \to X$ be a curve in X, defined on an interval J. We define α to be a **geodesic with respect to** F if the curve

$$\alpha': J \to TX$$

is an integral curve of F. Since $\pi\alpha' = \alpha$, that is α' lies above α in TX, we can express the geodesic condition equivalently by stating that α satisfies the relation

$$\alpha'' = F(\alpha').$$

This relation for curves α in X is called the **second-order differential equation** for the curve α, determined by F. Observe that by definition, if β is an integral curve of F in TX, then $\pi\beta$ is a geodesic for the second order vector field F.

Next we shall give the representation of the second order vector field and of the integral curves in a chart.

Representation in charts

Let U be open in the vector space $\mathbf{E}$, so that $T(U) = U \times \mathbf{E}$, and $T\big(T(U)\big) = (U \times \mathbf{E}) \times (\mathbf{E} \times \mathbf{E})$. Then $\pi: U \times \mathbf{E} \to U$ is simply the projection, and we have a commutative diagram:

$$
\begin{array}{ccc}
(U \times \mathbf{E}) \times (\mathbf{E} \times \mathbf{E}) & \xrightarrow{\;\pi_*\;} & U \times \mathbf{E} \\
\big\downarrow & & \big\downarrow \\
U \times \mathbf{E} & \xrightarrow[\pi]{} & U
\end{array}
$$

The map π_* on each fiber $\mathbf{E} \times \mathbf{E}$ is constant, and is simply the projection of $\mathbf{E} \times \mathbf{E}$ on the first factor $\mathbf{E}$, that is

$$\pi_*(x, v, u, w) = (x, u).$$

Any vector field on $U \times \mathbf{E}$ has a local representation

$$f: U \times \mathbf{E} \to \mathbf{E} \times \mathbf{E}$$

which has therefore two components, $f = (f_1, f_2)$, each f_i mapping $U \times \mathbf{E}$

into **E**. The next statement describes second order vector fields locally in the chart.

*Let U be open in the vector space **E**, and let $T(U) = U \times \mathbf{E}$ be the tangent bundle. A C^{p-2}-morphism*

$$f: \ U \times \mathbf{E} \to \mathbf{E} \times \mathbf{E}$$

is the local representation of a second order vector field on U if and only if

$$f(x, v) = \big(v, \ f_2(x, v)\big).$$

The above statement is merely making explicit the relation $\pi_* F = \mathrm{id}$, in the chart. If we write $f = (f_1, f_2)$, then we see that

$$f_1(x, v) = v.$$

We express the above relations in terms of integral curves as follows. Let $\beta = \beta(t)$ be an integral curve for the vector field F on TX. In the chart, the curve has two components

$$\beta(t) = \big(x(t), \ v(t)\big) \in U \times \mathbf{E}.$$

By definition, if f is the local representation of F, we must have

$$\frac{d\beta}{dt} = \left(\frac{dx}{dt}, \frac{dv}{dt}\right) = f(x, v) = \big(v, \ f_2(x, v)\big).$$

Consequently, our differential equation can be rewritten in the following manner:

$$\frac{dx}{dt} = v(t),$$

(1)
$$\frac{d^2 x}{dt^2} = \frac{dv}{dt} = f_2\left(x, \frac{dx}{dt}\right),$$

which is of course familiar.

Sprays

We shall be interested in special kinds of second-order differential equations. Before we discuss these, we make a few technical remarks.

Let s be a real number, and $\pi: \ E \to X$ be a vector bundle. If v is in E, so in E_x for some x in X, then sv is again in E_x since E_x is a vector space. We write s_E for the mapping of E into itself given by this scalar multiplication. This maping is in fact a VB-morphism, and even a VB-isomorphism if $s \neq 0$. Then

$$T(s_E) = (s_E)_*: \ T(E) \to T(E)$$

is the usual induced map on the tangent bundle of E.

Now let $E = TX$ be the tangent bundle itself. Then our map s_{TX} satisfies the property

$$(s_{TX})_* \circ s_{TTX} = s_{TTX} \circ (s_{TX})_*,$$

which follows from the linearity of s_{TX} on each fiber, and can also be seen directly from the representation on charts given below.

We define a **spray** to be a second-order vector field which satisfies the homogeneous quadratic condition:

SPR 1. *For all* $s \in \mathbf{R}$ *and* $v \in T(X)$, *we have*

$$F(sv) = (s_{TX})_* sF(v).$$

It is immediate from the conditions defining sprays (second-order vector field satisfying **SPR 1**) that **sprays form a convex set**! Hence if we can exhibit sprays over open subsets of vector spaces, then we can glue them together by means of partitions of unity, and we obtain at once the following global existence theorem.

Theorem 3.1. *Let X be a manifold of class C^p ($p \geq 3$). If X admits partitions of unity, then there exists a spray over X.*

Representations in a chart

Let U be open in $\mathbf{E}$, so that $TU = U \times \mathbf{E}$. Then

$$TTU = (U \times \mathbf{E}) \times (\mathbf{E} \times \mathbf{E}),$$

and the representations of s_{TU} and $(s_{TU})_*$ in the chart are given by the maps

$$s_{TU}: \ (x, v) \mapsto (x, sv) \qquad \text{and} \qquad (s_{TU})_*: \ (x, v, u, w) \mapsto (x, sv, u, sw).$$

Thus

$$s_{TTU} \circ (s_{TU})_*: \ (x, v, u, w) \mapsto (x, sv, su, s^2 w).$$

We may now give the local condition for a second-order vector field F to be a spray.

Proposition 3.2. *In a chart $U \times E$ for TX, let $f: U \times E \to E \times E$ represent F, with $f = (f_1, f_2)$. Then f represents a spray if and only if, for all $s \in \mathbf{R}$ we have*

$$f_2(x, sv) = s^2 f_2(x, v).$$

Proof. The proof follows at once from the definitions and the formula giving the chart representation of $s(s_{TX})_*$.

Thus we see that the condition **SPR 1** (in addition to being a second-order vector field), simply means that f_2 is homogeneous of degree 2 in the variable v. By the remark in Chapter I, §3, it follows that f_2 is a quadratic map in its second variable, and specifically, this quadratic map is given by

$$f_2(x, v) = \tfrac{1}{2} D_2^2 f_2(x, 0)(v, v).$$

Thus the spray is induced by a symmetric bilinear map given at each point x in a chart by

$$(2) \qquad\qquad B(x) = \tfrac{1}{2} D_2^2 f_2(x, 0).$$

Conversely, suppose given a morphism

$$U \to L_{\text{sym}}^2(\mathbf{E}, \mathbf{E}) \qquad \text{given by} \qquad x \mapsto B(x)$$

from U into the space of symmetric bilinear maps $\mathbf{E} \times \mathbf{E} \to \mathbf{E}$. Thus for each v, $w \in \mathbf{E}$ the value of $B(x)$ at (v, w) is denoted by $B(x; v, w)$ or $B(x)(v, w)$. Define $f_2(x, v) = B(x; v, v)$. Then f_2 is quadratic in its second variable, and the map f defined by

$$f(x, v) = \big(v, B(x; v, v)\big) = \big(v, f_2(x, v)\big)$$

represents a spray over U. We call B the **symmetric bilinear map associated with the spray**. From the local representations in (1) and (2), we conclude that *a curve α is a geodesic if and only if α satisfies the differential equation*

$$(3) \qquad\qquad \alpha''(t) = B_{\alpha(t)}\big(\alpha'(t), \alpha'(t)\big) \qquad \text{for all } t.$$

We recall the trivial fact from linear algebra that the bilinear map B is determined purely algebraically from the quadratic map, by the formula

$$B(v, w) = \tfrac{1}{2}[f_2(v + w) - f_2(v) - f_2(w)].$$

We have suppressed the x from the notation to focus on the relevant second variable v. Thus the quadratic map and the symmetric bilinear map determine each other uniquely.

The above discussion has been local, over an open set U in a Banach space. In Proposition 3.4 and the subsequent discussion of connections, we show how to globalize the bilinear map B intrinsically on the manifold.

Examples. As a trivial special case, we can always take $f_2(x, v) = (v, 0)$ to represent the second component of a spray in the chart.

In the chapter on Riemannian metrics, we shall see how to construct a spray in a natural fashion, depending on the metric.

In [La 99] the chapter on covariant derivatives, we show how a spray gives rise to such derivatives.

Next, let us give the transformation rule for a spray under a change of charts, i.e. an isomorphism $h\colon U \to V$. On TU, the map Th is represented by a morphism (its vector component)

$$H\colon U \times \mathbf{E} \to \mathbf{E} \times \mathbf{E} \qquad \text{given by} \qquad H(x, v) = \big(h(x), h'(x)v\big).$$

We then have one further lift to the double tangent bundle TTU, and we may represent the diagram of maps symbolically as follows:

$$
\begin{array}{ccc}
(U \times \mathbf{E}) \times (\mathbf{E} \times \mathbf{E}) & \xrightarrow{\ (H, H')\ } & (V \times \mathbf{E}) \times (\mathbf{E} \times \mathbf{E}) \\
\Big\downarrow \nearrow{\scriptstyle f_{U,2}} & & \Big\downarrow \nearrow{\scriptstyle f_{V,2}} \\
U \times \mathbf{E} & \xrightarrow{\ H = (h, h')\ } & V \times \mathbf{E} \\
\Big\downarrow & & \Big\downarrow \\
U & \xrightarrow{\ \ h\ \ } & V
\end{array}
$$

Then the derivative $H'(x, v)$ is given by the Jacobian matrix operating on column vectors ${}^t(u, w)$ with $u, w \in \mathbf{E}$, namely

$$H'(x, v) = \begin{pmatrix} h'(x) & 0 \\ h''(x)v & h'(x) \end{pmatrix} \quad \text{so } H'(x, v)\begin{pmatrix} u \\ w \end{pmatrix} = \begin{pmatrix} h'(x) & 0 \\ h''(x)v & h'(x) \end{pmatrix}\begin{pmatrix} u \\ w \end{pmatrix}.$$

Thus the top map on elements in the diagram is given by

$$(H, H')\colon (x, v, u, w) \mapsto \big(h(x), h'(x)v, h'(x)u, h''(x)(u, v) + h'(x)w\big).$$

For the application, we put $u = v$ because $f_1(x, v) = v$, and $w = f_{U,2}(x, v)$, where f_U and f_V denote the representations of the spray over U and V

respectively. It follows that f_U and f_V are related by the formula

$$f_V\big(h(x), h'(x)v\big) = \big(h'(x)v, h''(x)(v, v) + h'(x)f_{U,2}(x, v)\big).$$

Therefore we obtain:

Proposition 3.3. Change of variable formula for the quadratic part of a spray:

$$f_{V,2}\big(h(x), h'(x)v\big) = h''(x)(v, v) + h'(x)f_{U,2}(x, v),$$
$$B_V\big(h(x); h'(x)v, h'(x)w\big) = h''(x)(v, w) + h'(x)B_U(x; v, w).$$

Proposition 3.3 admits a converse:

Proposition 3.4. *Suppose we are given a covering of the manifold X by open sets corresponding to charts U, V,..., and for each U we are given a morphism*

$$B_U: \ U \to L^2_{\mathrm{sym}}(\mathbf{E}, \mathbf{E})$$

which transforms according to the formula of Proposition 3.3 under an isomorphism $h: U \to V$. Then there exists a unique spray whose associated bilinear map in the chart U is given by B_U.

Proof. We leave the verification to the reader.

Remarks. Note that $B_U(x; v, w)$ does not transform like a tensor of type $L^2_{\mathrm{sym}}(\mathbf{E}, \mathbf{E})$, i.e. a section of the bundle $L^2_{\mathrm{sym}}(TX, TX)$. There are several ways of defining the bilinear map B intrinsically. One of them is via second order bundles, or bundles of second order jets, and to extend the terminology we have established previously to such bundles, and even higher order jet bundles involving higher derivatives, as in [Po 62]. Another way is in [La 99], via connections. For our immediate purposes, it suffices to have the above discussion on second-order differential equations together with Proposition 3.3 and 3.4. Sprays were introduced by Ambrose, Palais, and Singer [APS 60], and I used them (as recommended by Palais) in the earliest version [La 62]. In [Lo 69] the bilinear map B_U is expressed in terms of second order jets. The basics of differential topology and geometry were being established in the early sixties. Cf. the bibliographical notes from [Lo 69] at the end of his first chapter.

Connections

We now show how to define the bilinear map B intrinsically and directly.

Matters will be clearer if we start with an arbitrary vector bundle

$$p\colon E \to X$$

over a manifold X. As it happens we also need the notion of a fiber bundle when the fibers are not necessarily vector spaces, so don't have a linear structure. Let $f\colon Y \to X$ be a morphism. We say that f (or Y over X) is a **fiber bundle** if f is surjective, and if each point x of X has an open neighborhood U, and there is some manifold Z and an isomorphism $h\colon f^{-1}(U) \to U \times Z$ such that the following diagram is commutative:

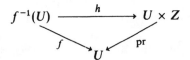

Thus locally, $f\colon Y \to X$ looks like the projection from a product space. The reason why we need a fiber bundle is that the tangent bundle

$$\pi_E\colon TE \to E$$

is a vector bundle over E, but the composite $f = p \circ \pi_E\colon TE \to X$ is only a fiber bundle over X, a fact which is obvious by picking trivializations in charts. Indeed, if U is a chart in X, and if $U \times \mathbf{F} \to U$ is a vector bundle chart for E, with fiber $\mathbf{F}$, and $Y = TE$, then we have a natural iso-morphism of fiber bundles over U:

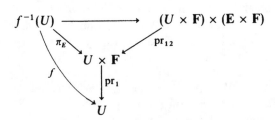

Note that U being a chart in X implies that $U \times \mathbf{E} \to U$ is a vector bundle chart for the tangent bundle TU over U.

The tangent bundle TE has two natural maps making it a vector bundle:

$$\pi_E\colon TE \to E \text{ is a vector bundle over } E;$$

$$T(p)\colon TE \to TX \text{ is a vector bundle over } TX.$$

Therefore we have a natural morphism of fiber bundle (not vector bundle) over X:

$$\big(\pi_E, T(p)\big): \ TE \to E \oplus TX \qquad \text{given by} \qquad W \mapsto \big(\pi_E W, T(p)W\big)$$

for $W \in TE$. If $W \in T_e E$ with $e \in E_x$, then $\pi_E W \in E_x$ and $T(p)W \in T_x X$.

After these preliminaries, we define a **connection** to be a morphism of fiber bundles over X, from the direct sum $E \oplus TX$ into TE:

$$H: \ E \oplus TX \to TE$$

such that

$$\big(\pi_E, T(p)\big) \circ H = \mathrm{id}_{E \oplus TX},$$

and such that H is bilinear, in other words $H_x: \ E_x \oplus T_x X \to TE$ is bilinear.

Consider a chart U as in the above diagram, so

$$TU = U \times \mathbf{E} \qquad \text{and} \qquad T(U \times \mathbf{F}) = (U \times \mathbf{F}) \times (\mathbf{E} \times \mathbf{F}).$$

Then our map H has a coordinate representation

$$H(x, e, v) = \big(x, e, H_1(x, e, v), H_2(x, e, v)\big) \qquad \text{for } e \in \mathbf{F} \text{ and } v \in \mathbf{E}.$$

The fact that $\big(\pi_E, T(p)\big) \circ H = \mathrm{id}_{E \oplus TX}$ implies at once that $H_1(x, e, v) = v$. The bilinearity condition implies that for fixed x, the map

$$(e, v) \mapsto H_2(x, e, v)$$

is bilinear as a map $\mathbf{F} \times \mathbf{E} \to \mathbf{E}$. We shall therefore denote this map by $B(x)$, and we write in the chart

$$H(x, e, v) = \big(x, e, v, B(x)(e, v)\big) \qquad \text{or also} \qquad \big(x, e, v, B(x, e, v)\big).$$

Now take the special case when $E = TX$. We say that the connection is **symmetric** if the bilinear map B is symmetric. Suppose this is the case. We may define the corresponding quadratic map $TX \to TTX$ by letting $f_2(x, v) = B(x, v, v)$. Globally, this amounts to defining a morphism

$$F: \ TX \to TTX \qquad \text{such that} \qquad F = H \circ \text{diagonal}$$

where the diagonal is taken in $TX \oplus TX$, in each fiber. Thus

$$F(v) = H(v, v) \qquad \text{for } v \in T_x X.$$

Then F is a vector field on TX, and the condition $(\pi_*, \pi_*) \circ H = \mathrm{id}$ on $TX \oplus TX$ implies that F is a second-order vector field on X, in other words, F defines a spray. It is obvious that all sprays can be obtained in this fashion. Thus we have shown how to describe geometrically the bilinear map associated with a spray.

Going back to the general case of a vector bundle E unrelated to TX, we note that the image of a connection H is a vector subbundle over E. Let V denote the kernel of the map $T(p): TE \to TX$. We leave it to the reader to verify in charts that V is a vector subbundle of TE over E, and that the image of H is a complementary subbundle. One calls V the **vertical subbundle**, canonically defined, and one calls H the **horizontal subbundle** determined by the connection. Cf. Kobayashi [Ko 57], Dombrowski [Do 68], and Besse [Be 78] for more basic material on connections.

IV, §4. THE FLOW OF A SPRAY AND THE EXPONENTIAL MAP

The condition we have taken to define a spray is equivalent to other conditions concerning the integral curves of the second-order vector field F. We shall list these conditions systematically. We shall use the following relation. If $\alpha: J \to X$ is a curve, and α_1 is the curve defined by $\alpha_1(t) = \alpha(st)$, then

$$\alpha_1'(t) = s\alpha'(st),$$

this being the chain rule for differentiation.

If v is a vector in TX, let β_v be the unique integral curve of F with initial condition v (i.e. such that $\beta_v(0) = v$). In the next three conditions, the sentence should begin with "for each v in TX".

SPR 2. *A number t is in the domain of β_{sv} if and only if st is in the domain of β_v and then*

$$\beta_{sv}(t) = s\beta_v(st).$$

SPR 3. *If s, t are numbers, st is in the domain of β_v if and only if s is in the domain of β_{tv}, and then*

$$\pi\beta_{tv}(s) = \pi\beta_v(st).$$

SPR 4. *A number t is in the domain of β_v if and only if 1 is in the domain of β_{tv}, and then*

$$\pi\beta_v(t) = \pi\beta_{tv}(1).$$

We shall now prove the equivalence between all four conditions.

Assume **SPR 1**, and let s be fixed. For all t such that st is in the domain of β_v, the curve $\beta_v(st)$ is defined and we have

$$\frac{d}{dt}\left(s\beta_v(st)\right) = s_* s\beta_v'(st) = s_* sF\left(\beta_v(st)\right) = F\left(s\beta_v(st)\right).$$

Hence the curve $s\beta_v(st)$ is an integral curve for F, with initial condition $s\beta_v(0) = sv$. By uniqueness we must have

$$s\beta_v(st) = \beta_{sv}(t).$$

This proves **SPR 2**.

Assume **SPR 2**. Since β_v is an integral curve of F for each v, with initial condition v, we have by definition

$$\beta_{sv}'(0) = F(sv).$$

Using our assumption, we also have

$$\beta_{sv}'(t) = \frac{d}{dt}\left(s\beta_v(st)\right) = s_* s\beta_v'(st).$$

Put $t = 0$. Then **SPR 1** follows because β_{sv} and β_v are integral curves of F with initial conditions sv and v respectively.

It is obvious that **SPR 2** implies **SPR 3**. Conversely, assume **SPR 3**. To prove **SPR 2**, we have

$$\beta_{sv}(t) = (\pi\beta_{sv})'(t) = \frac{d}{dt}\pi\beta_v(st) = s(\pi\beta_v)'(st) = s\beta_v(st),$$

which proves **SPR 2**.

Assume **SPR 4**. Then st is in the domain of β_r if and only if 1 is in the domain of β_{stv}, and s is in the domain of β_{tv} if and only if 1 is in the domain of β_{stv}. This proves the first assertion of **SPR 3**, and again by **SPR 4**, assuming these relations, we get **SPR 3**.

It is similarly clear that **SPR 3** implies **SPR 4**.

Next we consider further properties of the integral curves of a spray. Let F be a spray on X. As above, we let β_v be the integral curve with initial condition v. Let $\mathfrak{D}$ be the set of vectors v in $T(X)$ such that β_v is defined at least on the interval $[0, 1]$. We know from Corollary 2.7 that $\mathfrak{D}$ is an open set in $T(X)$, and by Theorem 2.6 the map

$$v \mapsto \beta_v(1)$$

is a morphism of $\mathfrak{D}$ into $T(X)$. We now define the **exponential** map

$$\exp: \mathfrak{D} \to X$$

to be

$$\exp(v) = \pi\beta_v(1).$$

Then exp is a C^{p-2}-morphism. We also call $\mathfrak{D}$ the **domain of the exponential map (associated with** F**)**.

If $x \in X$ and 0_x denotes the zero vector in T_x, then from **SPR 1**, taking $s = 0$, we see that $F(0_x) = 0$. Hence

$$\exp(0_x) = x.$$

Thus our exponential map coincides with π on the zero cross section, and so induces an isomorphism of the cross section onto X. It will be convenient to denote the zero cross section of a vector bundle E over X by $\zeta_E(X)$ or simply ζX if the reference to E is clear. Here, E is the tangent bundle.

We denote by $\exp_x$ the restriction of exp to the tangent space T_x. Thus

$$\exp_x: T_x \to X.$$

Theorem 4.1. *Let X be a manifold and F a spray on X. Then*

$$\exp_x: T_x \to X$$

induces a local isomorphism at 0_x, and in fact $(\exp_x)_*$ is the identity at 0_x.

Proof. We prove the second assertion first because the main assertion follows from it by the inverse mapping theorem. Furthermore, since T_x is a vector space, it suffices to determine the derivative of $\exp_x$ on rays, in other words, to determine the derivative with respect to t of a curve $\exp_x(tv)$. This is done by using **SPR 3**, and we find

$$\frac{d}{dt}\pi\beta_{tv} = \beta_{tv}.$$

Evaluating this at $t = 0$ and taking into account that β_w has w as initial condition for any w gives us

$$(\exp_x)_*(0_x) = \text{id}.$$

This concludes the proof of Theorem 4.1.

Helgason gave a general formula for the differential of the exponential map on analytic manifolds [He 61], reproduced in [He 78], Chapter I, Theorem 6.5.

Next we describe all geodesics.

Proposition 4.2. *The images of straight segments through the origin in T_x, under the exponential map $\exp_x$, are geodesics. In other words, if $v \in T_x$ and we let*

$$\alpha(v, t) = \alpha_v(t) = \exp_x(tv),$$

then α_v is a geodesic. Conversely, let $\alpha\colon J \to X$ be a C^2 geodesic defined on an interval J containing 0, and such that $\alpha(0) = x$. Let $\alpha'(0) = v$. Then $\alpha(t) = \exp_x(tv)$.

Proof. The first statement by definition means that α_v' is an integral curve of the spray F. Indeed, by the **SPR** conditions, we know that

$$\alpha(v, t) = \alpha_v(t) = \pi\beta_{tv}(1) = \pi\beta_v(t),$$

and $(\pi\beta_v)' = \beta_v$ is indeed an integral curve of the spray. Thus our assertion that the curves $t \mapsto \exp(tv)$ are geodesics is obvious from the definition of the exponential map and the **SPR** conditions.

Conversely, given a geodesic $\alpha\colon J \to X$, by definition α' satisfies the differential equation

$$\alpha''(t) = F(\alpha'(t)).$$

The two curves $t \mapsto \alpha(t)$ and $t \mapsto \exp_x(tv)$ satisfy the same differential equation and have the same initial conditions, so the two curves are equal. This proves the second statement and concludes the proof of the proposition.

Remark. From the theorem, we note that a C^1 curve in X is a geodesic if and only if, after a linear reparametrization of its interval of definition, it is simply $t \mapsto \exp_x(tv)$ for some x and some v.

We call the map $(v, t) \mapsto \alpha(v, t)$ the **geodesic flow** on X. It is defined on an open subset of $TX \times \mathbf{R}$, with $\alpha(v, 0) = x$ if $v \in T_x X$. Note that since $\pi(s\beta_v(t)) = \pi\beta_v(t)$ for $s \in \mathbf{R}$, we obtain from **SPR 2** the property

$$\alpha(sv, t) = \alpha(v, st)$$

for the geodesic flow. Precisely, t is in the domain of α_{sv} if and only if st is in the domain of α_v, and in that case the formula holds. As a slightly more precise version of Theorem 4.1 in this light, we obtain:

Corollary 4.3. *Let F be a spray on X, and let $x_0 \in X$. There exists an open neighborhood U of x_0, and an open neighborhood V of 0_{x_0} in TX satisfying the following condition. For every $x \in U$ and $v \in V \cap T_x X$, there exists a unique geodesic*

$$\alpha_v \colon (-2, 2) \to X$$

such that

$$\alpha_v(0) = x \qquad and \qquad \alpha_v'(0) = v.$$

Observe that in a chart, we may pick V as a product

$$V = U \times V_2(0) \subset U \times \mathbf{E}$$

where $V_2(0)$ is a neighborhood of 0 in $\mathbf{E}$. Then the geodesic flow is defined on $U \times V_2(0) \times J$, where $J = (-2, 2)$. We picked $(-2, 2)$ for concreteness. What we really want is that 0 and 1 lie in the interval. Any bounded interval J containing 0 and 1 could have been selected in the statement of the corollary. Then of course, U and V (or $V_2(0)$) depend on J.

IV, §5. EXISTENCE OF TUBULAR NEIGHBORHOODS

Let X be a submanifold of a manifold Y. A **tubular neighborhood** of X in Y consists of a vector bundle $\pi \colon E \to X$ over X, an open neighborhood Z of the zero section $\zeta_E X$ in E, and an isomorphism

$$f \colon Z \to U$$

of Z onto an open set in Y containing X, which commutes with ζ:

We shall call f the **tubular map** and Z or its image $f(Z)$ the corresponding **tube** (in E or Y respectively). The bottom map j is simply the inclusion. We could obviously assume that it is an embedding and define tubular neighborhoods for embeddings in the same way. We shall say that our tubular neighborhood is **total** if $Z = E$. In this section, we investigate conditions under which such neighborhoods exist. We shall consider the uniqueness problem in the next section.

Theorem 5.1. *Let Y be of class C^p $(p \geq 3)$ and admit partitions of unity. Let X be a closed submanifold. Then there exists a tubular neighborhood of X in Y, of class C^{p-2}.*

Proof. Consider the exact sequence of tangent bundles:

$$0 \to T(X) \to T(Y)|X \to N(X) \to 0.$$

We know that this sequence splits, and thus there exists some splitting

$$T(Y)|X = T(X) \oplus N(X)$$

where $N(X)$ may be identified with a subbundle of $T(Y)|X$. Following Palais, we construct a spray ξ on $T(Y)$ using Theorem 3.1 and obtain the corresponding exponential map. We shall use its restriction to $N(X)$, denoted by $\exp|N$. Thus

$$\exp|N: \, \mathfrak{D} \cap N(X) \to Y.$$

We contend that this map is a local isomorphism. To prove this, we may work locally. Corresponding to the submanifold, we have a product decomposition $U = U_1 \times U_2$, with $X = U_1 \times 0$. If U is open in $\mathbf{E}$, then we may take U_1, U_2 open in $\mathbf{F}_1$, $\mathbf{F}_2$ respectively. Then the injection of $N(X)$ in $T(Y)|X$ may be represented locally by an exact sequence

$$0 \to U_1 \times \mathbf{F}_2 \xrightarrow{\varphi} U_1 \times \mathbf{F}_1 \times \mathbf{F}_2,$$

and the inclusion of $T(Y)|X$ in $T(Y)$ is simply the inclusion

$$U_1 \times \mathbf{F}_1 \times \mathbf{F}_2 \to U_1 \times U_2 \times \mathbf{F}_1 \times \mathbf{F}_2.$$

We work at the point $(x_1, 0)$ in $U_1 \times \mathbf{F}_2$. We must compute the derivative of the composite map

$$U_1 \times \mathbf{F}_2 \xrightarrow{\varphi} U_1 \times U_2 \times \mathbf{F}_1 \times \mathbf{F}_2 \xrightarrow{\exp} Y$$

at $(x_1, 0)$. We can do this by the formula for the partial derivatives. Since the exponential map coincides with the projection on the zero cross section, its "horizontal" partial derivative is the identity. By Theorem 4.1 we know that its "vertical" derivative is also the identity. Let

$$\psi = (\exp) \circ \bar{\varphi}$$

(where $\bar{\varphi}$ is simply φ followed by the inclusion). Then for any vector (w_1, w_2) in $\mathbf{F}_1 \times \mathbf{F}_2$ we get

$$D\psi(x_1, 0) \cdot (w_1, w_2) = (w_1, 0) + \varphi_{x_1}(w_2),$$

where φ_{x_1} is the linear map given by φ on the fiber over x_1. By hypothesis, we know that $\mathbf{F}_1 \times \mathbf{F}_2$ is the direct sum of $\mathbf{F}_1 \times 0$ and of the image of φ_{x_1}. This proves that $D\psi(x_1, 0)$ is a toplinear isomorphism, and in fact proves that **the exponential map restricted to a normal bundle is a local isomorphism** on the zero cross section.

We have thus shown that there exists a vector bundle $E \to X$, an open neighborhood Z of the zero section in E, and a mapping $f: Z \to Y$ which, for each x in ζ_E, is a local isomorphism at x. We must show that Z can be shrunk so that f restricts to an isomorphism. To do this we follow Godement ([God 58], p. 150). We can find a locally finite open covering of X by open sets U_i in Y such that, for each i we have inverse isomorphisms

$$f_i: Z_i \to U_i \qquad \text{and} \qquad g_i: U_i \to Z_i$$

between U_i and open sets Z_i in Z, such that each Z_i contains a point x of X, such that f_i, g_i are the identity on X (viewed as a subset of both Z and Y) and such that f_i is the restriction of f to Z_i. We now find a locally finite covering $\{V_i\}$ of X by open sets of Y such that $\overline{V}_i \subset U_i$, and let $V = \bigcup V_i$. We let W be the subset of elements $y \in V$ such that, if y lies in an intersection $\overline{V}_i \cap \overline{V}_j$, then $g_i(y) = g_j(y)$. Then W certainly contains X. We contend that W contains an open subset containing X.

Let $x \in X$. There exists an open neighborhood G_x of x in Y which *meets only a finite number of* $\overline{V}_i$, say $\overline{V}_{i_1}, \dots, \overline{V}_{i_r}$. *Taking* G_x *small enough, we can assume that* x lies in each one of these, and that G_x is contained in each one of the sets $\overline{U}_{i_1}, \dots, \overline{U}_{i_r}$. Since x lies in each $\overline{V}_{i_1}, \dots, \overline{V}_{i_r}$, it is contained in $U_{i_1}, \dots, U_{i_r}$ and our maps $g_{i_1}, \dots, g_{i_r}$ take the same value at x, namely x itself. Using the fact that $f_{i_1}, \dots, f_{i_r}$ are restrictions of f, we see at once that our finite number of maps $g_{i_1}, \dots, g_{i_r}$ must agree on G_x if we take G_x small enough.

Let G be the union of the G_x. Then G is open, and we can define a map

$$g: G \to g(G) \subset Z$$

by taking g equal to g_i on $G \cap V_i$. Then $g(G)$ is open in Z, and the restriction of f to $g(G)$ is an inverse for g. This proves that f, g are inverse isomorphisms on G and $g(G)$, and concludes the proof of the theorem.

A vector bundle $E \to X$ will be said to be **compressible** if, given an open neighborhood Z of the zero section, there exists an isomorphism

$$\varphi: E \to Z_1$$

of E with an open subset Z_1 of Z containing the zero section, which commutes with the projection on X:

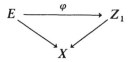

It is clear that if a bundle is compressible, and if we have a tubular neighborhood defined on Z, then we can get a total tubular neighborhood defined on E. We shall see in the chapter on Riemannian metrics that certain types of vector bundles are compressible.

IV, §6. UNIQUENESS OF TUBULAR NEIGHBORHOODS

Let X, Y be two manifolds, and $F: \mathbf{R} \times X \to Y$ a morphism. We shall say that F is an **isotopy** (of embeddings) if it satisfies the following conditions. First, for each $t \in \mathbf{R}$, the map F_t given by $F_t(x) = F(t, x)$ is an embedding. Second, there exist numbers $t_0 < t_1$ such that $F_t = F_{t_0}$ for all $t \leq t_0$ and $F_{t_1} = F_t$ for all $t \geq t_1$. We then say that the interval $[t_0, t_1]$ is a **proper domain** for the isotopy, and the constant embeddings on the left and right will also be denoted by $F_{-\infty}$ and $F_{+\infty}$ respectively. We say that two embeddings $f: X \to Y$ and $g: X \to Y$ are **isotopic** if there exists an isotopy F_t as above such that $f = F_{t_0}$ and $g = F_{t_1}$ (notation as above). We write $f \approx g$ for f isotopic to g.

Using translations of intervals, and multiplication by scalars, we can always transform an isotopy to a new one whose proper domain is contained in the interval $(0, 1)$. Furthermore, the relation of isotopy between embeddings is an equivalence relation. It is obviously symmetric and reflexive, and for transitivity, suppose $f \approx g$ and $g \approx h$. We can choose the ranges of these isotopies so that the first one ends and stays constant at g before the second starts moving. Thus it is clear how to compose isotopies in this case.

If $s_0 < s_1$ are two numbers, and $\sigma: \mathbf{R} \to \mathbf{R}$ is a function (morphism) such that $\sigma(s) = t_0$ for $s \leq s_0$ and $\sigma(s) = t_1$ for $s \geq s_1$, and σ is monotone increasing, then from a given isotopy F_t we obtain another one, $G_t = F_{\sigma(t)}$. Such a function σ can be used to smooth out a piece of isotopy given only on a closed interval.

Remark. We shall frequently use the following trivial fact: If $f_t: X \to Y$ is an isotopy, and if $g: X_1 \to X$ and $h: Y \to Y_1$ are two embeddings, then the composite map

$$hf_tg: X_1 \to Y_1$$

is also an isotopy.

Let Y be a manifold and X a submanifold. Let $\pi\colon E \to X$ be a vector bundle, and Z an open neighborhood of the zero section. An isotopy $f_t\colon Z \to Y$ of open embeddings such that each f_t is a tubular neighborhood of X will be called an **isotopy of tubular neighborhoods**. In what follows, the domain will usually be all of E.

Proposition 6.1. *Let X be a manifold. Let $\pi\colon E \to X$ and $\pi_1\colon E_1 \to X$ be two vector bundles over X. Let*

$$f\colon E \to E_1$$

be a tubular neighborhood of X in E_1 (identifying X with its zero section in E_1). Then there exists an isotopy

$$f_t\colon E \to E_1$$

with proper domain $[0, 1]$ such that $f_1 = f$ and f_0 is a VB-isomorphism. (If f, π, π_1 are of class C^p then f_t can be chosen of class C^{p-1}.)

Proof. We define F by the formula

$$F_t(e) = t^{-1} f(te)$$

for $t \neq 0$ and $e \in E$. Then F_t is an embedding since it is composed of embeddings (the scalar multiplications by t, t^{-1} are in fact VB-isomorphism).

We must investigate what happens at $t = 0$.

Given $e \in E$, we find an open neighborhood U_1 of πe over which E_1 admits a trivialization $U_1 \times \mathbf{E}_1$. We then find a still smaller open neighborhood U of πe and an open ball B around 0 in the typical fiber $\mathbf{E}$ of E such that E admits a trivialization $U \times \mathbf{E}$ over U, and such that the representation $\bar{f}$ of f on $U \times B$ (contained in $U \times \mathbf{E}$) maps $U \times B$ into $U_1 \times \mathbf{E}_1$. This is possible by continuity. On $U \times B$ we can represent $\bar{f}$ by two morphisms,

$$\bar{f}(x, v) = (\varphi(x, v),\, \psi(x, v))$$

and $\varphi(x, 0) = x$ while $\psi(x, 0) = 0$. Observe that for all t sufficiently small, te is contained in $U \times B$ (in the local representation).

We can represent F_t locally on $U \times B$ as the mapping

$$\bar{F}_t(x, v) = (\varphi(x, tv),\, t^{-1}\psi(x, tv)).$$

The map φ is then a morphism in the three variables x, v, and t even at

$t = 0$. The second component of $\bar{F}_t$ can be written

$$t^{-1}\psi(x, tv) = t^{-1}\int_0^1 D_2\psi(x, stv) \cdot (tv) \, ds$$

and thus t^{-1} cancels t to yield simply

$$\int_0^1 D_2\psi(x, stv) \cdot v \, ds.$$

This is a morphism in t, even at $t = 0$. Furthermore, for $t = 0$, we obtain

$$\bar{F}_0(x, v) = \big(x, D_2\psi(x, 0)v\big).$$

Since f was originally assumed to be an embedding, it follows that $D_2\psi(x, 0)$ is a toplinear isomorphism, and therefore F_0 is a VB-isomorphism. To get our isotopy in standard form, we can use a function $\sigma: \mathbf{R} \to \mathbf{R}$ such that $\sigma(t) = 0$ for $t \leq 0$ and $\sigma(t) = 1$ for $t \geq 1$, and σ is monotone increasing. This proves our proposition.

Theorem 6.2. *Let X be a submanifold of Y. Let*

$$\pi: E \to X \qquad and \qquad \pi_1: E_1 \to X$$

be two vector bundles, and assume that E is compressible. Let $f: E \to Y$ and $g: E_1 \to Y$ be two tubular neighborhoods of X in Y. Then there exists a C^{p-1}-isotopy

$$f_t: E \to Y$$

of tubular neighborhoods with proper domain $[0, 1]$ and a VB-isomorphism $\lambda: E \to E_1$ such that $f_1 = f$ and $f_0 = g\lambda$.

Proof. We observe that $f(E)$ and $g(E_1)$ are open neighborhoods of X in Y. Let $U = f^{-1}\big(f(E) \cap g(E_1)\big)$ and let $\varphi: E \to U$ be a compression. Let ψ be the composite map

$$E \xrightarrow{\varphi} U \xrightarrow{f|U} Y$$

$\psi = (f|U) \circ \varphi$. Then ψ is a tubular neighborhood, and $\psi(E)$ is contained in $g(E_1)$. Therefore $g^{-1}\psi: E \to E_1$ is a tubular neighborhood of the same type considered in the previous proposition. There exists an isotopy of tubular neighborhoods of X:

$$G_t: E \to E_1$$

such that $G_1 = g^{-1}\psi$ and G_0 is a VB-isomorphism. Considering the isotopy gG_t, we find an isotopy of tubular neighborhoods

$$\psi_t: E \to Y$$

such that $\psi_1 = \psi$ and $\psi_0 = g\omega$ where $\omega: E \to E_1$ is a VB-isomorphism. We have thus shown that ψ and $g\omega$ are isotopic (by an isotopy of tubular neighborhoods). Similarly, we see that ψ and $f\mu$ are isotopic for some VB-isomorphism

$$\mu: E \to E.$$

Consequently, adjusting the proper domains of our isotopies suitably, we get an isotopy of tubular neighborhoods going from $g\omega$ to $f\mu$, say F_t. Then $F_t\mu^{-1}$ will give us the desired isotopy from $g\omega\mu^{-1}$ to f, and we can put $\lambda = \omega\mu^{-1}$ to conclude the proof.

(By the way, the uniqueness proof did not use the existence theorem for differential equations.)

CHAPTER V

Operations on Vector Fields and Differential Forms

If $E \to X$ is a vector bundle, then it is of considerable interest to investigate the special operation derived from the functor "multilinear alternating forms." Applying it to the tangent bundle, we call the sections of our new bundle differential forms. One can define formally certain relations between functions, vector fields, and differential forms which lie at the foundations of differential and Riemannian geometry. We shall give the basic system surrounding such forms. In order to have at least one application, we discuss the fundamental 2-form, and in the next chapter connect it with Riemannian metrics in order to construct canonically the spray associated with such a metric.

We assume throughout that our manifolds are sufficiently differentiable so that all of our statements make sense.

V, §1. VECTOR FIELDS, DIFFERENTIAL OPERATORS, BRACKETS

Let X be a manifold of class C^p and φ a function defined on an open set U, that is a morphism

$$\varphi: U \to \mathbf{R}.$$

Let ξ be a vector field of class C^{p-1}. Recall that

$$T_x\varphi: T_x(U) \to T_x(\mathbf{R}) = \mathbf{R}$$

is a linear map. With it, we shall define a new function to be denoted by $\xi\varphi$ or $\xi \cdot \varphi$, or $\xi(\varphi)$. (There will be no confusion with this notation and composition of mappings.)

Proposition 1.1. *There exists a unique function $\xi\varphi$ on U of class C^{p-1} such that*

$$(\xi\varphi)(x) = (T_x\varphi)\xi(x).$$

If U is open in the vector space $\mathbf{E}$ and ξ denotes the local representation of the vector field on U, then

$$(\xi\varphi)(x) = \varphi'(x)\xi(x).$$

Proof. The first formula certainly defines a mapping of U into $\mathbf{R}$. The local formula defines a C^{p-1}-morphism on U. It follows at once from the definitions that the first formula expresses invariantly in terms of the tangent bundle the same mapping as the second. Thus it allows us to define $\xi\varphi$ as a morphism globally, as desired.

Let Fu^p denote the ring of functions (of class C^p). Then our operation $\varphi \mapsto \xi\varphi$ gives rise to a linear map

$$\partial_\xi \colon \mathrm{Fu}^p(U) \to \mathrm{Fu}^{p-1}(U), \qquad \text{defined by} \quad \partial_\xi\varphi = \xi\varphi.$$

A mapping

$$\partial \colon R \to S$$

from a ring R into an R-algebra S is called a **derivation** if it satisfies the usual formalism: Linearity, and $\partial(ab) = a\partial(b) + \partial(a)b$.

Proposition 1.2. *Let X be a manifold and U open in X. Let ξ be a vector field over X. If $\partial_\xi = 0$, then $\xi(x) = 0$ for all $x \in U$. Each ∂_ξ is a derivation of $\mathrm{Fu}^p(U)$ into $\mathrm{Fu}^{p-1}(U)$.*

Proof. Suppose $\xi(x) \neq 0$ for some x. We work with the local representations, and take φ to be a linear map of $\mathbf{E}$ into $\mathbf{R}$ such that $\varphi(\xi(x)) \neq 0$. Then $\varphi'(y) = \varphi$ for all $y \in U$, and we see that $\varphi'(x)\xi(x) \neq 0$, thus proving the first assertion. The second is obvious from the local formula.

From Proposition 1.2 we deduce that if two vector fields induce the same differential operator on the functions, then they are equal.

Given two vector fields ξ, η on X, we shall now define a new vector field $[\xi, \eta]$, called their **bracket product**.

Proposition 1.3. *Let ξ, η be two vector fields of class C^{p-1} on X. Then there exists a unique vector field $[\xi, \eta]$ of class C^{p-2} such that for each open set U and function φ on U we have*

$$[\xi, \eta]\varphi = \xi(\eta(\varphi)) - \eta(\xi(\varphi)).$$

If U is open in **E** *and* ξ, η *are the local representations of the vector fields, then* $[\xi, \eta]$ *is given by the local formula*

$$[\xi, \eta]\varphi(x) = \varphi'(x)\big(\eta'(x)\xi(x) - \xi'(x)\eta(x)\big).$$

Thus the local representation of $[\xi, \eta]$ *is given by*

$$[\xi, \eta](x) = \eta'(x)\xi(x) - \xi'(x)\eta(x).$$

Proof. By Proposition 1.2, any vector field having the desired effect on functions is uniquely determined. We check that the local formula gives us this effect locally. Differentiating formally, we have (using the law for the derivative of a product):

$$(\eta\varphi)'\xi - (\xi\varphi)'\eta = (\varphi'\eta)'\xi - (\varphi'\xi)\eta$$
$$= \varphi'\eta'\xi + \varphi''\eta\xi - \varphi'\xi'\eta - \varphi''\xi\eta.$$

The terms involving φ'' must be understood correctly. For instance, the first such term at a point x is simply $\varphi''(x)\big(\eta(x), \xi(x)\big)$ remembering that $\varphi''(x)$ is a bilinear map, and can thus be evaluated at the two vectors $\eta(x)$ and $\xi(x)$. However, we know that $\varphi''(x)$ is symmetric. Hence the two terms involving the second derivative of φ cancel, and give us our formula.

Corollary 1.4. *The bracket* $[\xi, \eta]$ *is bilinear in both arguments, we have* $[\xi, \eta] = -[\eta, \xi]$, *and Jacobi's identity*

$$[\xi, [\eta, \zeta]] = [[\xi, \eta], \zeta] + [\eta, [\xi, \zeta]].$$

In other words, for each ξ *the map* $\eta \mapsto [\xi, \eta]$ *is a derivation with respect to the Lie product* $(\eta, \zeta) \mapsto [\eta, \zeta]$.
If φ *is a function, then*

$$[\xi, \varphi\eta] = (\xi\varphi)\eta + \varphi[\xi, \eta], \qquad and \qquad [\varphi\xi, \eta] = \varphi[\xi, \eta] - (\eta\varphi)\xi.$$

Proof. The first two assertions are obvious. The third comes from the definition of the bracket. We apply the vector field on the left of the equality to a function φ. All the terms cancel out (the reader will write it out as well or better than the author). The last two formulas are immediate.

We make some comments concerning the functoriality of vector fields. Let

$$f : X \to Y$$

be an isomorphism. Let ξ be a vector field over X. Then we obtain an induced vector field $f_*\xi$ over Y, defined by the formula

$$(f_*\xi)(f(x)) = Tf(\xi(x)).$$

It is the vector field making the following diagram commutative:

$$
\begin{array}{ccc}
TX & \xrightarrow{\ Tf\ } & TY \\
\xi \uparrow\downarrow & & \downarrow\uparrow f_*\xi \\
X & \xrightarrow{\ f\ } & Y
\end{array}
$$

We shall also write f^* for $(f^{-1})_*$ when applied to a vector field. Thus we have the formulas

$$\boxed{f_*\xi = Tf \circ \xi \circ f^{-1}} \quad \text{and} \quad \boxed{f^*\xi = Tf^{-1} \circ \xi \circ f.}$$

If f is not an isomorphism, then one cannot in general define the direct or inverse image of a vector field as done above. However, let ξ be a vector field over X, and let η be a vector field over Y. If for each $x \in X$ we have

$$Tf(\xi(x)) = \eta(f(x)),$$

then we shall say that f maps ξ into η, or that ξ and η are f-**related**. If this is the case, then we may denote by $f_*\xi$ the map from $f(X)$ into TY defined by the above formula.

Let ξ_1, ξ_2 be vector fields over X, and let η_1, η_2 be vector fields over Y. If ξ_i is f-related to η_i for $i = 1, 2$ then as maps on $f(X)$ we have

$$\boxed{f_*[\xi_1, \xi_2] = [\eta_1, \eta_2].}$$

We may write suggestively the formula in the form

$$\boxed{f_*[\xi_1, \xi_2] = [f_*\xi_1, f_*\xi_2].}$$

Of course, this is meaningless in general, since $f_*\xi_1$ may not be a vector field on Y. When f is an isomorphism, then it is a correct formulation of the other formula. In any case, it suggests the correct formula.

To prove the formula, we work with the local representations, when $X = U$ is open in $\mathbf{E}$, and $Y = V$ is open in $\mathbf{F}$. Then ξ_i, η_i are maps of U, V into the spaces $\mathbf{E}$, $\mathbf{F}$ respectively. For $x \in X$ we have

$$(f_*[\xi_1, \xi_2])(x) = f'(x)(\xi_2'(x)\xi_1(x) - \xi_1'(x)\xi_2(x)).$$

On the other hand, by assumption, we have

$$\eta_i(f(x)) = f'(x)\xi_i(x),$$

so that

$$
\begin{aligned}
[\eta_1, \eta_2](f(x)) &= \eta_2'(f(x))\eta_1(f(x)) - \eta_1'(f(x))\eta_2(f(x)) \\
&= \eta_2'(f(x))f'(x)\xi_1(x) - \eta_1'(f(x))f'(x)\xi_2(x) \\
&= (\eta_2 \circ f)'(x)\xi_1(x) - (\eta_1 \circ f)'(x)\xi_2(x) \\
&= f''(x) \cdot \xi_2(x) \cdot \xi_1(x) + f'(x)\xi_2'(x)\xi_1(x) \\
&\quad - f''(x) \cdot \xi_1(x) \cdot \xi_2(x) - f'(x)\xi_1'(x)\xi_2(x).
\end{aligned}
$$

Since $f''(x)$ is symmetric, two terms cancel, and the remaining two terms give the same value as $(f_*[\xi_1, \xi_2])(x)$, as was to be shown.

The bracket between vector fields gives an infinitesimal criterion for commutativity in various contexts. We give here one theorem of a general nature as an example of this phenomenon.

Theorem 1.5. *Let ξ, η be vector fields on X, and assume that $[\xi, \eta] = 0$. Let α and β be the flows for ξ and η respectively. Then for real values t, s we have*

$$\alpha_t \circ \beta_s = \beta_s \circ \alpha_t.$$

Or in other words, for any $x \in X$ we have

$$\alpha(t, \beta(s, x)) = \beta(s, \alpha(t, x)),$$

in the sense that if for some value of t a value of s is in the domain of one of these expressions, then it is in the domain of the other and the two expressions are equal.

Proof. For a fixed value of t, the two curves in s given by the right- and left-hand side of the last formula have the same initial condition, namely $\alpha_t(x)$. The curve on the right

$$s \mapsto \beta(s, \alpha(t, x))$$

is by definition the integral curve of η. The curve on the left

$$s \mapsto \alpha\big(t, \beta(s, x)\big)$$

is the image under α_t of the integral curve for η having initial condition x. Since x is fixed, let us denote $\beta(s, x)$ simply by $\beta(s)$. What we must show is that the two curves on the right and on the left satisfy the same differential equation.

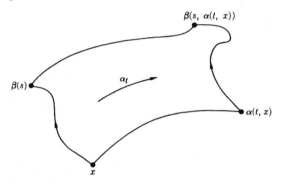

In the above figure, we see that the flow α_t shoves the curve on the left to the curve on the right. We must compute the tangent vectors to the curve on the right. We have

$$\frac{d}{ds}\big(\alpha_t(\beta(s))\big) = D_2\alpha\big(t, \beta(s)\big)\beta'(s)$$

$$= D_2\alpha\big(t, \beta(s)\big)\eta\big(\beta(s)\big).$$

Now fix s, and denote this last expression by $F(t)$. We must show that if

$$G(t) = \eta\big(\alpha(t, \beta(s))\big),$$

then

$$F(t) = G(t).$$

We have trivially $F(0) = G(0)$, in other words the curves F and G have the same initial condition. On the other hand,

$$F'(t) = \xi'\big(\alpha(t, \beta(s))\big) D_2\alpha\big(t, \beta(s)\big)\eta\big(\beta(s)\big)$$

and

$$G'(t) = \eta'\big(\alpha(t, \beta(s))\big)\xi\big(\alpha(t, \beta(s))\big)$$

$$= \xi'\big(\alpha(t, \beta(s))\big)\eta\big(\alpha(t, \beta(s))\big) \quad \text{(because } [\xi, \eta] = 0\text{).}$$

Hence we see that our two curves F and G satisfy the same differential equation, whence they are equal. This proves our theorem.

Vector fields ξ, η such that $[\xi, \eta] = 0$ are said to **commute**. One can generalize the process of straightening out vector fields to a finite number of commuting vector fields, using the same method of proof, using Theorem 1.5. As another application, one can prove that if the Lie algebra of a connected Lie group is commutative, then the group is commutative. Cf. the section on Lie groups.

V, §2. LIE DERIVATIVE

Let λ be a differentiable functor on vector spaces. For convenience, take λ to be covariant and in one variable. What we shall say in the rest of this section would hold in the same way (with slightly more involved notation) if λ had several variables and were covariant in some and contravariant in others.

Given a manifold X, we can take $\lambda(T(X))$. It is a vector bundle over X, which we denote by $T_\lambda(X)$ as in Chapter III. Its sections $\Gamma_\lambda(X)$ are the tensor fields of type λ.

Let ξ be a vector field on X, and U open in X. It is then possible to associate with ξ a map

$$\mathscr{L}_\xi \colon \Gamma_\lambda(U) \to \Gamma_\lambda(U)$$

(with a loss of two derivatives). This is done as follows.

Given a point x of U and a local flow α for ξ at x, we have for each t sufficiently small a local isomorphism α_t in a neighborhood of our point x. Recall that locally, $\alpha_t^{-1} = \alpha_{-t}$. If η is a tensor field of type λ, then the composite mapping $\eta \circ \alpha_t$ has its range in $T_\lambda(X)$. Finally, we can take the tangent map $T(\alpha_{-t}) = (\alpha_{-t})_*$ to return to $T_\lambda(X)$ in the fiber above x. We thus obtain a composite map

$$F(t, x) = (\alpha_{-t})_* \circ \eta \circ \alpha_t(x) = (\alpha_t^* \eta)(x),$$

which is a morphism, locally at x. We take its derivative with respect to t and evaluate it at 0. After looking at the situation locally in a trivialization of $T(X)$ and $T_\lambda(X)$ at x, one sees that the map one obtains gives a section of $T_\lambda(U)$, that is a tensor field of type λ over U. This is our map $\mathscr{L}_\xi$. To summarize,

$$\mathscr{L}_\xi \eta = \frac{d}{dt}\bigg|_{t=0} (\alpha_{-t})_* \circ \eta \circ \alpha_t.$$

This map $\mathscr{L}_\xi$ is called the **Lie derivative**. We shall determine the Lie derivative on functions and on vector fields in terms of notions already discussed.

First let φ be a function. Then by the general definition, the Lie derivative of this function with respect to the vector field ξ with flow α is defined to be

$$\mathscr{L}_\xi \varphi(x) = \lim_{t \to 0} \frac{1}{t} \big[\varphi(\alpha(t,\,x)) - \varphi(x)\big],$$

or in other words,

$$\mathscr{L}_\xi \varphi = \frac{d}{dt} (\alpha_t^* \varphi)\Big|_{t=0}.$$

Our assertion is then that

$$\boxed{\mathscr{L}_\xi \varphi = \xi\varphi.}$$

To prove this, let

$$F(t) = \varphi(\alpha(t,\,x)).$$

Then

$$F'(t) = \varphi'(\alpha(t,\,x))\,D_1\alpha(t,\,x)$$
$$= \varphi'(\alpha(t,\,x))\,\xi(\alpha(t,\,x)),$$

because α is a flow for ξ. Using the initial condition at $t = 0$, we find that

$$F'(0) = \varphi'(x)\xi(x),$$

which is precisely the value of $\xi\varphi$ at x, thus proving our assertion.

If ξ, η are vector fields, then

$$\boxed{\mathscr{L}_\xi \eta = [\xi,\,\eta].}$$

As before, let α be a flow for ξ. The Lie derivative is given by

$$\mathscr{L}_\xi \eta = \frac{d}{dt} (\alpha_t^* \eta)\Big|_{t=0}.$$

Letting ξ and η denote the local representations of the vector fields, we note that the local representation of $(\alpha_t^* \eta)(x)$ is given by

$$(\alpha_t^* \eta)(x) = F(t) = D_2\alpha(-t,\,x)\eta(\alpha(t,\,x)).$$

We must therefore compute $F'(t)$, and then $F'(0)$. Using the chain rule, the formula for the derivative of a product, and the differential equation satisfied by $D_2\alpha$, we obtain

$$F'(t) = -D_1 D_2\alpha(-t, x)\eta\big(\alpha(t, x)\big) + D_2\alpha(-t, x)\eta'\big(\alpha(t, x)\big)D_1\alpha(t, x)$$
$$= -\xi'\big(\alpha(-t, x)\big)D_2\alpha(-t, x)\eta\big(\alpha(t, x)\big) + D_2\alpha(-t, x)\eta'\big(\alpha(t, x)\big).$$

Putting $t = 0$ proves our formula, taking into account the initial conditions

$$\alpha(0, x) = x \qquad \text{and} \qquad D_2\alpha(0, x) = \text{id}.$$

V, §3. EXTERIOR DERIVATIVE

Let X be a manifold. The functor L_a^r (r-multilinear continuous alternating forms) extends to arbitrary vector bundles, and in particular, to the tangent bundle of X. A **differential form** of degree r, or simply an r-**form** on X, is a section of $L_a^r(T(X))$, that is a tensor field of type L_a^r. If X is of class C^p, forms will be assumed to be of a suitable class C^s with $1 \leq s \leq p - 1$. The set of differential forms of degree r will be denoted by $\mathscr{A}^r(X)$ ($\mathscr{A}$ for alternating). It is not only a vector space (infinite dimensional) over $\mathbf{R}$ but a module over the ring of functions on X (of the appropriate order of differentiability). If ω is an r-form, then $\omega(x)$ is an element of $L_a^r(T_x(X))$, and is thus an r-multilinear alternating form of $T_x(X)$ into $\mathbf{R}$. We sometimes denote $\omega(x)$ by ω_x.

Suppose U is open in the vector space $\mathbf{E}$. Then $L_a^r(T(U))$ is equal to $U \times L_a^r(\mathbf{E})$ and a differential form is entirely described by the projection on the second factor, which we call its **local representation**, following our general system (Chapter III, §4). Such a local representation is therefore a morphism

$$\omega\colon\; U \to L_a^r(\mathbf{E}).$$

Let ω be in $L_a^r(\mathbf{E})$ and $v_1, \ldots, v_r$ elements of $\mathbf{E}$. We denote the value $\omega(v_1, \ldots, v_r)$ also by

$$\langle \omega, v_1 \times \cdots \times v_r \rangle.$$

Similarly, let $\xi_1, \ldots, \xi_r$ be vector fields on an open set U, and let ω be an r-form on X. We denote by

$$\langle \omega, \xi_1 \times \cdots \times \xi_r \rangle$$

the mapping from U into $\mathbf{R}$ whose value at a point x in U is

$$\langle \omega(x), \xi_1(x) \times \cdots \times \xi_r(x) \rangle.$$

Looking at the situation locally on an open set U such that $T(U)$ is trivial, we see at once that this mapping is a morphism (i.e. a function on U) of the same degree of differentiability as ω and the ξ_i.

Proposition 3.1. *Let x_0 be a point of X and ω an r-form on X. If*

$$\langle \omega, \xi_1 \times \cdots \times \xi_r \rangle(x_0)$$

is equal to 0 for all vector fields $\xi_1, \ldots, \xi_r$ at x_0 (i.e. defined on some neighborhood of x_0), then $\omega(x_0) = 0$.

Proof. Considering things locally in terms of their local representations, we see that if $\omega(x_0)$ is not 0, then it does not vanish at some r-tuple of vectors $(v_1, \ldots, v_r)$. We can take vector fields at x_0 which take on these values at x_0 and from this our assertion is obvious.

It is convenient to agree that a differential form of degree 0 is a function. In the next proposition, we describe the exterior derivative of an r-form, and it is convenient to describe this situation separately in the case of functions.

Therefore let $f: X \to \mathbf{R}$ be a function. For each $x \in X$, the tangent map

$$T_x f: \ T_x(X) \to T_{f(x)}(\mathbf{R}) = \mathbf{R}$$

is a continuous linear map, and looking at local representations shows at once that the collection of such maps defines a 1-form which will be denoted by df. Furthermore, from the definition of the operation of vector fields on functions, it is clear that df is the unique 1-form such that for every vector field ξ we have

$$\langle df, \xi \rangle = \xi f.$$

To extend the definition of d to forms of higher degree, we recall that if

$$\omega: \ U \to L_a^r(\mathbf{E})$$

is the local representation of an r-form over an open set U of $\mathbf{E}$, then for each x in U,

$$\omega'(x): \ \mathbf{E} \to L_a^r(\mathbf{E})$$

is a continuous linear map. Applied to a vector v in $\mathbf{E}$, it therefore gives rise to an r-form on $\mathbf{E}$.

Proposition 3.2. *Let ω be an r-form of class C^{p-1} on X. Then there exists a unique $(r+1)$-form $d\omega$ on X of class C^{p-2} such that, for any* *open set U of X and vector fields $\xi_0, \ldots, \xi_r$ on U we have*

$$\langle d\omega, \xi_0 \times \cdots \times \xi_r \rangle$$

$$= \sum_{i=0}^{r} (-1)^i \xi_i \langle \omega, \xi_0 \times \cdots \times \hat{\xi}_i \times \cdots \times \xi_r \rangle$$

$$+ \sum_{i<j} (-1)^{i+j} \langle \omega, [\xi_i, \xi_j] \times \xi_0 \times \cdots \times \hat{\xi}_i \times \cdots \times \hat{\xi}_j \times \cdots \times \xi_r \rangle.$$

If furthermore U is open in **E** *and ω, $\xi_0, \ldots, \xi_r$ are the local representations of the form and the vector fields respectively, then at a point x the value of the expression above is equal to*

$$\sum_{i=0}^{r} (-1)^i \langle \omega'(x)\xi_i(x), \xi_0(x) \times \cdots \times \widehat{\xi_i(x)} \times \cdots \times \xi_r(x) \rangle.$$

Proof. As before, we observe that the local formula defines a differential form. If we can prove that it gives the same thing as the first formulas, which is expressed invariantly, then we can globalize it, and we are done. Let us denote by S_1 and S_2 the two sums occurring in the invariant expression, and let L be the local expression. We must show that $S_1 + S_2 = L$. We consider S_1, and apply the definition of ξ_i operating on a function locally, as in Proposition 1.1, at a point x. We obtain

$$S_1 = \sum_{i=0}^{r} (-1)^i \langle \omega, \xi_0 \times \cdots \times \hat{\xi}_i \times \cdots \times \xi_r \rangle'(x)\xi_i(x).$$

The derivative is perhaps best computed by going back to the definition. Applying this definition directly, and discarding second order terms, we find that S_1 is equal to

$$\sum (-1)^i \langle \omega'(x)\xi_i(x), \xi_0(x) \times \cdots \times \widehat{\xi_i(x)} \times \cdots \times \xi_r(x) \rangle$$

$$+ \sum_i \sum_{i<j} (-1)^i \langle \omega(x), \xi_0(x) \times \cdots \times \xi_j'(x)\xi_i(x) \times \cdots \times \widehat{\xi_i(x)} \times \cdots \times \xi_r(x) \rangle$$

$$+ \sum_i \sum_{j<i} \langle \omega(x), \xi_0(x) \times \cdots \times \widehat{\xi_i(x)} \times \cdots \times \xi_j'(x)\xi_i(x) \times \cdots \times \xi_r(x) \rangle.$$

Of these there sums, the first one is the local formula L. As for the other two, permuting j and i in the first, and moving the term $\xi_i'(x)\xi_i(x)$ to the first position, we see that they combine to give (symbolically)

$$- \sum_i \sum_{i<j} (-1)^{i+j} \langle \omega, (\xi_j'\xi_i - \xi_i'\xi_j) \times \xi_0 \times \cdots \times \hat{\xi}_i \times \cdots \times \hat{\xi}_j \times \cdots \times \xi_r \rangle$$

(evaluated at x). Using Proposition 1.3, we see that this combination is equal to $-S_2$. This proves that $S_1 + S_2 = L$, as desired.

We call $d\omega$ the **exterior derivative** of ω. Leaving out the order of differentiability for simplicity, we see that d is an **R**-linear map

$$d\colon \mathscr{A}r(X) \to \mathscr{A}^{r+1}(X).$$

We now look into the multiplicative properties of d with respect to the wedge product.

Let ω, ψ be multilinear alternating forms of degree r and s respectively on the vector space **E**. In multilinear algebra, one defines their **wedge product** as an $(r + s)$ multilinear alternating form, by the formula

$$(\omega \wedge \psi)(v_1, \ldots, v_{r+s}) = \frac{1}{r!\, s!} \sum \epsilon(\sigma) \omega(v_{\sigma 1}, \ldots, v_{\sigma r}) \psi(v_{\sigma(r+1)}, \ldots, v_{\sigma(r+s)})$$

the sum being taken over all permutations σ of $(1, \ldots, r + s)$. This definition extends at once to differential forms on a manifold, if we view it as giving the value for $\omega \wedge \psi$ at a point x. The v_i are then elements of the tangent space T_x, and considering the local representations shows at once that the wedge product so defined gives a morphism of the manifold X into $L_a^{r+s}(T(X))$, and is therefore a differential form.

Remark. The coefficient $1/r!\, s!$ is not universally taken to define the wedge product. Some people, e.g. [He 78] and [KoN 63], take $1/(r + s)!$, which causes constants to appear later. I have taken the same factor as [AbM 78] and [GHL 87/93]. I recommend that the reader check out the case with $r = s = 1$ so $r + s = 2$ to see how a factor $\frac{1}{2}$ comes in. With either convention, the wedge product between forms is associative, so with some care, one can carry out a consistent theory with either convention. I leave the proof of associativity to the reader. It follows by induction that if $\omega_1, \ldots, \omega_m$ are forms of degrees $r_1, \ldots, r_m$ respectively, and $r = r_1 + \cdots + r_m$, then

$$(\omega_1 \wedge \cdots \wedge \omega_m)(v_1, \ldots, v_r) = \frac{1}{r_1! \cdots r_m!} \sum_\sigma \epsilon(\sigma) \Omega_\sigma,$$

where

$$\Omega_\sigma = \omega_1(v_{\sigma 1}, \ldots, v_{\sigma r_1}) \omega_2(v_{\sigma(r_1+1)}, \ldots, v_{\sigma(r_1+r_2)}) \cdots \omega_m(v_{\sigma(r-r_m+1)}, \ldots, v_{\sigma r}),$$

and where the sum is taken over all permutations of $(1, \ldots, r)$.

If we regard functions on X as differential forms of degree 0, then the ordinary product of a function by a differential form can be viewed as the wedge product. Thus if f is a function and ω a differential form, then

$$f\omega = f \wedge \omega.$$

(The form on the left has the value $f(x)\omega(x)$ at x.)

The next proposition gives us more formulas concerning differential forms.

Proposition 3.3. *Let* ω, ψ *be differential forms on* X. *Then*

EXD 1. $d(\omega \wedge \psi) = d\omega \wedge \psi + (-1)^{\deg(\omega)}\omega \wedge d\psi.$

EXD 2. $dd\omega = 0$ (*with enough differentiability, say* $p \geq 4$).

Proof. This is a simple formal exercise in the use of the local formula for the local representation of the exterior derivative. We leave it to the reader.

One can give a local representation for differential forms and the exterior derivative in terms of local coordinates, which are especially useful in integration which fits the notation better. We shall therefore carry out this local formulation in full. It dates back to Cartan [Ca 28]. There is in addition a theoretical point which needs clarifying. We shall use at first the wedge $\wedge$ in two senses. One sense is defined as above, giving rise to Proposition 3.3. Another sense will come from Theorem A. We shall comment on their relation after Theorem B.

We recall first two simple results from linear (or rather multilinear) algebra. We use the notation $\mathbf{E}^{(r)} = \mathbf{E} \times \mathbf{E} \times \cdots \times \mathbf{E}$, r times.

Theorem A. *Let* $\mathbf{E}$ *be a vector space over the reals of dimension* n. *For each positive integer* r *with* $1 \leq r \leq n$ *there exists a vector space* $\bigwedge^r \mathbf{E}$ *and a multilinear alternating map*

$$\mathbf{E}^{(r)} \rightarrow \bigwedge^r \mathbf{E}$$

denoted by $(u_1, \ldots, u_r) \mapsto u_1 \wedge \cdots \wedge u_r$, *having the following property: If* $\{v_1, \ldots, v_n\}$ *is a basis of* $\mathbf{E}$, *then the elements*

$$\{v_{i_1} \wedge \cdots \wedge v_{i_r}\}, \qquad i_1 < i_2 < \cdots < i_r,$$

form a basis of $\bigwedge^r \mathbf{E}$.

We recall that **alternating** means that $u_1 \wedge \cdots \wedge u_r = 0$ if $u_i = u_j$ for some $i \neq j$. We call $\bigwedge^r \mathbf{E}$ the r-th **alternating** product (or **exterior** product) on $\mathbf{E}$. If $r = 0$, we define $\bigwedge^0 \mathbf{E} = \mathbf{R}$. Elements of $\bigwedge^r \mathbf{E}$ which can be written in the form $u_1 \wedge \cdots \wedge u_r$ are called **decomposable**. Such elements generate $\bigwedge^r \mathbf{E}$. If $r > \dim E$, we define $\bigwedge^r \mathbf{E} = \{0\}$.

Theorem B. *For each pair of positive integers (r, s), there exists a unique product (bilinear map)*

$$\bigwedge^r \mathbf{E} \times \bigwedge^s \mathbf{E} \to \bigwedge^{r+s} \mathbf{E}$$

such that if $u_1, \ldots, u_r,\ w_1, \ldots, w_s \in \mathbf{E}$ then

$$(u_1 \wedge \cdots \wedge u_r) \times (w_1 \wedge \cdots \wedge w_s) \mapsto u_1 \wedge \cdots \wedge u_r \wedge w_1 \wedge \cdots \wedge w_s.$$

This product is associative.

The proofs for these two statements can be found, for instance, in my *Linear Algebra*.

Let $\mathbf{E}^\vee$ be the dual space, $\mathbf{E}^\vee = L(\mathbf{E}, \mathbf{R})$. If $\mathbf{E} = \mathbf{R}^n$ and $\lambda_1, \ldots, \lambda_n$ are the coordinate functions, then each λ_i is an element of the dual space, and in fact $\{\lambda_1, \ldots, \lambda_n\}$ is a basis of this dual space. Let $\mathbf{E} = \mathbf{R}^n$. There is a linear isomorphism

$$\boxed{\bigwedge^r \mathbf{E}^\vee \xrightarrow{\approx} L_a^r(\mathbf{E}, \mathbf{R})}$$

given in the following manner. If $g_1, \ldots, g_r \in \mathbf{E}^\vee$ and $v_1, \ldots, v_r \in \mathbf{E}$, then the value

$$\det\big(g_i(v_j)\big)$$

is multilinear alternating both as a function of $(g_1, \ldots, g_r)$ and $(v_1, \ldots, v_r)$. Thus it induces a pairing

$$\bigwedge^r \mathbf{E}^\vee \times \mathbf{E}^r \to \mathbf{R}$$

and a map

$$\bigwedge^r \mathbf{E}^\vee \to L_a^r(\mathbf{E}, \mathbf{R}).$$

This map is the isomorphism mentioned above. Using bases, it is easy to verify that it is an isomorphism (at the level of elementary algebra).

Thus in the finite dimensional case, we may identify $L_a^r(\mathbf{E}, \mathbf{R})$ with the alternating product $\bigwedge^r \mathbf{E}^\vee$, and consequently we may view the local representation of a differential form of degree r to be a map

$$\omega \colon U \to \bigwedge^r \mathbf{E}^\vee$$

from U into the rth alternating product of $\mathbf{E}^\vee$. We say that the form is of class C^p if the map is of class C^p. (We view $\bigwedge^r \mathbf{E}^\vee$ as a normed vector space, using any norm. It does not matter which, since all norms on a finite dimensional vector space are equivalent.) The wedge product as we gave it is compatible with the wedge product and the isomorphism of $\bigwedge^r \mathbf{E}$ with $L_a^r(\mathbf{E}, \mathbf{R})$ given above. If we had taken a different convention for the wedge product of alternating forms, then a constant would have appeared in front of the above determinant to establish the above identification (e.g. the constant $\frac{1}{2}$ in the 2×2 case).

Since $\{\lambda_1, \ldots, \lambda_n\}$ is a basis of $\mathbf{E}^\vee$, we can express each differential form in terms of its coordinate functions with respect to the basis

$$\{\lambda_{i_1} \wedge \cdots \wedge \lambda_{i_r}\}, \qquad (i_1 < \cdots < i_r),$$

namely for each $x \in U$ we have

$$\omega(x) = \sum_{(i)} f_{i_1 \cdots i_r}(x) \lambda_{i_1} \wedge \cdots \wedge \lambda_{i_r},$$

where $f_{(i)} = f_{i_1 \cdots i_r}$ is a function on U. Each such function has the same order of differentiability as ω. We call the preceding expression the **standard form** of ω. We say that a form is **decomposable** if it can be written as just one term $f(x) \lambda_{i_1} \wedge \cdots \wedge \lambda_{i_r}$. Every differential form is a sum of decomposable ones.

We agree to the convention that functions are differential forms of degree 0.

As before, the differential forms on U of given degree r form a vector space, denoted by $\mathscr{A}^r(U)$.

Let $\mathbf{E} = \mathbf{R}^n$. Let f be a function on U. For each $x \in U$ the derivative

$$f'(x) \colon \mathbf{R}^n \to \mathbf{R}$$

is a linear map, and thus an element of the dual space. Thus

$$f' \colon U \to \mathbf{E}^\vee$$

represents a differential form of degree 1, which is usually denoted by df. If f is of class C^p, then df is class C^{p-1}.

Let λ_i be the i-th coordinate function. Then we know that

$$d\lambda_i(x) = \lambda_i'(x) = \lambda_i$$

for each $x \in U$ because $\lambda'(x) = \lambda$ for any linear map λ. Whenever $\{x_1, \ldots, x_n\}$ are used systematically for the coordinates of a point in $\mathbf{R}^n$, it is customary in the literature to use the notation

$$d\lambda_i(x) = dx_i.$$

This is slightly incorrect, but is useful in formal computations. We shall also use it in this book on occasions. Similarly, we also write (incorrectly)

$$\omega = \sum_{(i)} f_{(i)} \, dx_{i_1} \wedge \cdots \wedge dx_{i_r}$$

instead of the correct

$$\omega(x) = \sum_{(i)} f_{(i)}(x)\lambda_{i_1} \wedge \cdots \wedge \lambda_{i_r}.$$

In terms of coordinates, the map df (or f') is given by

$$df(x) = f'(x) = D_1 f(x)\lambda_1 + \cdots + D_n f(x)\lambda_n,$$

where $D_i f(x) = \partial f / \partial x_i$ is the i-th partial derivative. This is simply a restatement of the fact that if $h = (h_1, \ldots, h_n)$ is a vector, then

$$f'(x)h = \frac{\partial f}{\partial x_1} h_1 + \cdots + \frac{\partial f}{\partial x_n} h_n.$$

Thus in old notation, we have

$$df(x) = \frac{\partial f}{\partial x_1} \, dx_1 + \cdots + \frac{\partial f}{\partial x_n} \, dx_n.$$

We shall develop the theory of the alternating product and the exterior derivative directly without assuming Propositions 3.2 or 3.3.

Let ω and ψ be forms of degrees r and s respectively, on the open set U. For each $x \in U$ we can then take the alternating product $\omega(x) \wedge \psi(x)$ and we define the **alternating product** $\omega \wedge \psi$ by

$$(\omega \wedge \psi)(x) = \omega(x) \wedge \psi(x).$$

(It is an exercise to verify that this product corresponds to the product defined previously before Proposition 3.3 under the isomorphism between $L_a^r(\mathbf{E}, \mathbf{R})$ and the r-th alternating product.) If f is a differential form of degree 0, that is a function, then we have again

$$f \wedge \omega = f\omega,$$

where $(f\omega)(x) = f(x)\omega(x)$. By definition, we then have

$$\omega \wedge f\psi = f\omega \wedge \psi.$$

We shall now define the **exterior derivative** $d\omega$ for any differential form ω. We have already done it for functions. We shall do it in general first in terms of coordinates, and then show that there is a characterization independent of these coordinates. If

$$\omega = \sum_{(i)} f_{(i)} \, d\lambda_{i_1} \wedge \cdots \wedge d\lambda_{i_r},$$

we define

$$d\omega = \sum_{(i)} df_{(i)} \wedge d\lambda_{i_1} \wedge \cdots \wedge d\lambda_{i_r}.$$

Example. Suppose $n = 2$ and ω is a 1-form, given in terms of the two coordinates (x, y) by

$$\omega(x, y) = f(x, y) \, dx + g(x, y) \, dy.$$

Then

$$
\begin{aligned}
d\omega(x, y) &= df(x, y) \wedge dx + dg(x, y) \wedge dy \\
&= \left(\frac{\partial f}{\partial x} \, dx + \frac{\partial f}{\partial y} \, dy \right) \wedge dx + \left(\frac{\partial g}{\partial x} \, dx + \frac{\partial g}{\partial y} \, dy \right) \wedge dy \\
&= \frac{\partial f}{\partial y} \, dy \wedge dx + \frac{\partial g}{\partial x} \, dx \wedge dy \\
&= \left(\frac{\partial f}{\partial y} - \frac{\partial g}{\partial x} \right) dy \wedge dx
\end{aligned}
$$

because the terms involving $dx \wedge dx$ and $dy \wedge dy$ are equal to 0.

Proposition 3.4. *The map d is linear, and satisfies*

$$d(\omega \wedge \psi) = d\omega \wedge \psi + (-1)^r \omega \wedge d\psi$$

if $r = \deg \omega$. The map d is uniquely determined by these properties, and by the fact that for a function f, we have $df = f'$.

Proof. The linearity of d is obvious. Hence it suffices to prove the formula for decomposable forms. We note that for any function f we have

$$d(f\omega) = df \wedge \omega + f \, d\omega.$$

Indeed, if ω is a function g, then from the derivative of a product we get $d(fg) = f \, dg + g \, df$. If

$$\omega = g \, d\lambda_{i_1} \wedge \cdots \wedge d\lambda_{i_r},$$

where g is a function, then

$$d(f\omega) = d(fg\, d\lambda_{i_1} \wedge \cdots \wedge d\lambda_{i_r}) = d(fg) \wedge d\lambda_{i_1} \wedge \cdots \wedge d\lambda_{i_r}$$
$$= (f\, dg + g\, df) \wedge d\lambda_{i_1} \wedge \cdots \wedge d\lambda_{i_r}$$
$$= f\, d\omega + df \wedge \omega,$$

as desired. Now suppose that

$$\omega = f\, d\lambda_{i_1} \wedge \cdots \wedge d\lambda_{i_r} \qquad \text{and} \qquad \psi = g\, d\lambda_{j_1} \wedge \cdots \wedge d\lambda_{j_s}$$
$$= f\tilde{\omega}, \qquad\qquad\qquad\qquad\qquad = g\tilde{\psi},$$

with $i_1 < \cdots < i_r$ and $j_1 < \cdots < j_s$ as usual. If some $i_v = j_\mu$, then from the definitions we see that the expressions on both sides of the equality in the theorem are equal to 0. Hence we may assume that the sets of indices $i_i, \ldots, i_r$ and $j_1, \ldots, j_s$ have no element in common. Then $d(\tilde{\omega} \wedge \tilde{\psi}) = 0$ by definition, and

$$d(\omega \wedge \psi) = d(fg\tilde{\omega} \wedge \tilde{\psi}) = d(fg) \wedge \tilde{\omega} \wedge \tilde{\psi}$$
$$= (g\, df + f\, dg) \wedge \tilde{\omega} \wedge \tilde{\psi}$$
$$= d\omega \wedge \psi + f\, dg \wedge \tilde{\omega} \wedge \tilde{\psi}$$
$$= d\omega \wedge \psi + (-1)^r f\tilde{\omega} \wedge dg \wedge \tilde{\psi}$$
$$= d\omega \wedge \psi + (-1)^r \omega \wedge d\psi,$$

thus proving the desired formula, in the present case. (We used the fact that $dg \wedge \tilde{\omega} = (-1)^r \tilde{\omega} \wedge dg$ whose proof is left to the reader.) The formula in the general case follows because any differential form can be expressed as a sum of forms of the type just considered, and one can then use the bilinearity of the product. Finally, d is uniquely determined by the formula, and its effect on functions, because any differential form is a sum of forms of type $f\, d\lambda_i \wedge \cdots \wedge d\lambda_{i_r}$ and the formula gives an expression of d in terms of its effect on forms of lower degree. By induction, if the value of d on functions is known, its value can then be determined on forms of degree ≥ 1. This proves our assertion.

Proposition 3.5. *Let ω be a form of class C^2. Then $dd\omega = 0$.*

Proof. If f is a function, then

$$df(x) = \sum_{j=1}^{n} \frac{\partial f}{\partial x_j}\, dx_j$$

and

$$ddf(x) = \sum_{j=1}^{n} \sum_{k=1}^{n} \frac{\partial^2 f}{\partial x_k \partial x_j} \, dx_k \wedge dx_j.$$

Using the fact that the partials commute, and the fact that for any two positive integers r, s we have $dx_r \wedge dx_s = -dx_s \wedge dx_r$, we see that the preceding double sum is equal to 0. A similar argument shows that the theorem is true for 1-forms, of type $g(x)\, dx_i$ where g is a function, and thus for all 1-forms by linearity. We proceed by induction. It suffices to prove the formula in general for decomposable forms. Let ω be decomposable of degree r, and write

$$\omega = \eta \wedge \psi,$$

where $\deg \psi = 1$. Using the formula for the derivative of an alternating product twice, and the fact that $dd\psi = 0$ and $dd\eta = 0$ by induction, we see at once that $dd\omega = 0$, as was to be shown.

We conclude this section by giving some properties of the pull-back of forms. As we saw at the end of Chapter III, §4, if $f: X \to Y$ is a morphism and if ω is a differential form on Y, then we get a differential form $f^*(\omega)$ on X, which is given at a point $x \in X$ by the formula

$$f^*(\omega)_x = \omega_{f(x)} \circ (T_x f)^r,$$

if ω is of degree r. This holds for $r \geq 1$. The corresponding local representation formula reads

$$\langle f^* \omega(x), \xi_1(x) \times \cdots \times \xi_r(x) \rangle = \langle \omega(f(x)), f'(x)\xi_1(x) \times \cdots \times f'(x)\xi_r(x) \rangle$$

if $\xi_1, \ldots, \xi_r$ are vector fields.

In the case of a 0-form, that is a function, its pull-back is simply the composite function. In other words, if φ is a function on Y, viewed as a form of degree 0, then

$$\boxed{f^*(\varphi) = \varphi \circ f.}$$

It is clear that the pull-back is linear, and satisfies the following properties.

Property 1. *If ω, ψ are two differential forms on Y, then*

$$f^*(\omega \wedge \psi) = f^*(\omega) \wedge f^*(\psi).$$

Property 2. *If ω is a differential form on Y, then*

$$df^*(\omega) = f^*(d\omega).$$

Property 3. *If* $f: X \to Y$ *and* $g: Y \to Z$ *are two morphisms, and* ω *is a differential form on* Z, *then*

$$f^*(g^*(\omega)) = (g \circ f)^*(\omega).$$

Finally, in the case of forms of degree 0:

Property 4. *If* $f: X \to Y$ *is a morphism, and* g *is a function on* Y, *then*

$$d(g \circ f) = f^*(dg)$$

and at a point $x \in X$, *the value of this* 1-*form is given by*

$$T_{f(x)}g \circ T_x f = (dg)_x \circ T_x f.$$

The verifications are all easy, and even trivial, except possibly for **Property 2**. We shall give the proof of **Property 2**.

For a form of degree 1, say

$$\omega(y) = g(y)\, dy_1,$$

with $y_1 = f_1(x)$, we find

$$(f^* d\omega)(x) = \left(g'(f(x)) \circ f'(x)\right) \wedge df_1(x).$$

Using the fact that $dd f_1 = 0$, together with Proposition 3.4 we get

$$(df^* \omega)(x) = \left(d(g \circ f)\right)(x) \wedge df_1(x),$$

which is equal to the preceding expression. Any 1-form can be expressed as a linear combination of form $g_i\, dy_i$, so that our assertion is proved for forms of degree 1.

The general formula can now be proved by induction. Using the linearity of f^*, we may assume that ω is expressed as $\omega = \psi \wedge \eta$ where ψ, η have lower degree. We apply Proposition 3.3 and **Property 1** to

$$f^* d\omega = f^*(d\psi \wedge \eta) + (-1)^r f^*(\psi \wedge d\eta)$$

and we see at once that this is equal to $df^*\omega$, because by induction, $f^* d\psi = df^*\psi$ and $f^* d\eta = df^*\eta$. This proves **Property 2**.

Example 1. Let $y_1, \ldots, y_m$ be the coordinates on V, and let μ_j be the jth coordinate function, $j = 1, \ldots, m$, so that $y_j = \mu_j(y_1, \ldots, y_m)$. Let

$$f: U \to V$$

be the map with coordinate functions

$$y_j = f_j(x) = \mu_j \circ f(x).$$

If

$$\omega(y) = g(y) \, dy_{j_1} \wedge \cdots \wedge dy_{j_s}$$

is a differential form on V, then

$$\boxed{f^*\omega = (g \circ f) \, df_{j_1} \wedge \cdots \wedge df_{j_s}.}$$

Indeed, we have for $x \in U$:

$$(f^*\omega)(x) = g(f(x))(\mu_{j_1} \circ f'(x)) \wedge \cdots \wedge (\mu_{j_s} \circ f'(x))$$

and

$$f_j'(x) = (\mu_j \circ f)'(x) = \mu_j \circ f'(x) = df_j(x).$$

Example 2. Let $f: [a, b] \to \mathbf{R}^2$ be a map from an interval into the plane, and let x, y be the coordinates of the plane. Let t be the coordinate in $[a, b]$. A differential form in the plane can be written in the form

$$\omega(x, y) = g(x, y) \, dx + h(x, y) \, dy,$$

where g, h are functions. Then by definition,

$$f^*\omega(t) = g(x(t), y(t)) \frac{dx}{dt} \, dt + h(x(t), y(t)) \frac{dy}{dt} \, dt,$$

if we write $f(t) = (x(t), y(t))$. Let $G = (g, h)$ be the vector field whose components are g and h. Then we can write

$$f^*\omega(t) = G(f(t)) \cdot f'(t) \, dt,$$

which is essentially the expression which is integrated when defining the integral of a vector field along a curve.

Example 3. Let U, V be both open sets in n-space, and let $f: U \to V$ be a C^p map. If

$$\omega(y) = g(y) \, dy_1 \wedge \cdots \wedge dy_n,$$

where $y_j = f_j(x)$ is the j-th coordinate of y, then

$$dy_j = D_1 f_j(x) \, dx_1 + \cdots + D_n f_j(x) \, dx_n$$

$$dy_j = \frac{\partial y_j}{\partial x_1} \, dx_1 + \cdots + \frac{\partial y_j}{\partial x_n} \, dx_n,$$

and consequently, expanding out the alternating product according to the usual multilinear and alternating rules, we find that

$$f^*\omega(x) = g(f(x))\Delta_f(x)\,dx_1 \wedge \cdots \wedge dx_n,$$

where Δ_f is the determinant of the Jacobian matrix of f.

V, §4. THE POINCARÉ LEMMA

If ω is a differential form on a manifold and is such that $d\omega = 0$, then it is customary to say that ω is **closed**. If there exists a form ψ such that $\omega = d\psi$, then one says that ω is **exact**. We shall now prove that locally, every closed form is exact.

Theorem 4.1 (Poincaré Lemma). *Let U be an open ball in $\mathbf{E}$ and let ω be a differential form of degree ≥ 1 on U such that $d\omega = 0$. Then there exists a differential form ψ on U such that $d\psi = \omega$.*

Proof. We shall construct a linear map k from the r-forms to the $(r-1)$-forms $(r \geq 1)$ such that

$$dk + kd = \text{id}.$$

From this relation, it will follow that whenever $d\omega = 0$, then

$$dk\omega = \omega,$$

thereby proving our proposition. We may assume that the center of the ball is the origin. If ω is an r-form, then we define $k\omega$ by the formula

$$\langle (k\omega)_x,\, v_1 \times \cdots \times v_{r-1} \rangle = \int_0^1 t^{r-1} \langle \omega(tx),\, x \times v_1 \times \cdots \times v_{r-1} \rangle\, dt.$$

We can assume that we deal with local representations and that $v_i \in \mathbf{E}$. We have

$$\langle (dk\omega)_x,\, v_1 \times \cdots \times v_r \rangle$$

$$= \sum_{i=1}^r (-1)^{i+1} \langle (k\omega)'(x)v_i,\, v_1 \times \cdots \times \hat{v}_i \times \cdots \times v_r \rangle$$

$$= \sum (-1)^{i+1} \int_0^1 t^{r-1} \langle \omega(tx),\, v_i \times v_1 \times \cdots \times \hat{v}_i \times \cdots \times v_r \rangle\, dt$$

$$+ \sum (-1)^{i+1} \int_0^1 t^r \langle \omega'(tx)v_i,\, x \times v_1 \times \cdots \times \hat{v}_i \times \cdots \times v_r \rangle\, dt.$$

On the other hand, we also have

$$\langle (kd\omega)(x), v_1 \times \cdots \times v_r \rangle$$

$$= \int_0^1 t^r \langle d\omega(x), x \times v_1 \times \cdots \times v_r \rangle \, dt$$

$$= \int_0^1 t^r \langle \omega'(tx)x, v_1 \times \cdots \times v_r \rangle \, dt$$

$$+ \sum (-1)^i \int_0^1 t^r \langle \omega'(tx)v_i, x \times v_1 \times \cdots \times \hat{v}_i \times \cdots \times v_r \rangle \, dt.$$

We observe that the second terms in the expressions for $kd\omega$ and $dk\omega$ occur with opposite signs and cancel when we take the sum. As to the first terms, if we shift v_i to the i-th place in the expression for $dk\omega$, then we get an extra coefficient of $(-1)^{i+1}$. Thus

$$dk\omega + kd\omega = \int_0^1 rt^{r-1} \langle \omega(tx), v_1 \times \cdots \times v_r \rangle \, dt$$

$$+ \int_0^1 t^r \langle \omega'(tx)x, v_1 \times \cdots \times v_r \rangle \, dt.$$

This last integral is simply the integral of the derivative with respect to t of

$$\langle t^r \omega(tx), v_1 \times \cdots \times v_r \rangle.$$

Evaluating this expression between $t = 0$ and $t = 1$ yields

$$\langle \omega(x), v_1 \times \cdots \times v_r \rangle$$

which proves the theorem.

We observe that we could have taken our open set U to be star-shaped instead of an open ball.

V, §5. CONTRACTIONS AND LIE DERIVATIVE

Let ξ be a vector field and let ω be an r-form on a manifold X, $r \geq 1$. Then we can define an $(r-1)$-form $C_\xi \omega$ by the formula

$$(C_\xi \omega)(x)(v_2, \ldots, v_r) = \omega(\xi(x), v_2, \ldots, v_r),$$

for $v_2, \ldots, v_r \in T_x$. Using local representations shows at once that $C_\xi \omega$ has the appropriate order of differentiability (the minimum of ω and ξ). We call $C_\xi \omega$ the **contraction** of ω by ξ, and also denote $C_\xi \omega$ by

$$\omega \circ \xi.$$

If f is a function, we define $C_\xi f = 0$. Leaving out the order of differentiability, we see that contraction gives an **R**-linear map

$$C_\xi \colon \mathscr{A}^r(X) \to \mathscr{A}^{r-1}(X).$$

This operation of contraction satisfies the following properties.

CON 1. $C_\xi \circ C_\xi = 0$.

CON 2. *The association* $(\xi, \omega) \mapsto C_\xi \omega = \omega \circ \xi$ *is bilinear. It is in fact bilinear with respect to functions, that is if φ is a function, then*

$$C_{\varphi \xi} = \varphi C_\xi \qquad and \qquad C_\xi(\varphi \omega) = \varphi C_\xi \omega.$$

CON 3. *If ω, ψ are differential forms and $r = \deg \omega$, then*

$$C_\xi(\omega \wedge \psi) = (C_\xi \omega) \wedge \psi + (-1)^r \omega \wedge C_\xi \psi.$$

These three properties follow at once from the definitions.

Example. Let $X = \mathbf{R}^n$, and let

$$\omega(x) = dx_1 \wedge \cdots \wedge dx_n.$$

If ξ is a vector field on $\mathbf{R}^n$, then we have the local representation

$$(\omega \circ \xi)(x) = \sum_{i=1}^n (-1)^{i+1} \xi_i(x) \, dx_1 \wedge \cdots \wedge \widehat{dx_i} \wedge \cdots \wedge dx_n.$$

We also have immediately from the definition of the exterior derivative,

$$d(\omega \circ \xi) = \sum_{i=1}^n \frac{\partial \xi_i(x)}{\partial x_i} \, dx_1 \wedge \cdots \wedge dx_n,$$

letting $\xi = (\xi_1, \ldots, \xi_n)$ in terms of its components ξ_i.

We can define the **Lie derivative** of an r-form as we did before for vector fields. Namely, we shall evaluate the following limit:

$$(\mathscr{L}_\xi \omega)(x) = \lim_{t \to 0} \frac{1}{t} [(\alpha_t^* \omega)(x) - \omega(x)],$$

or in other words,

$$\mathscr{L}_\xi \omega = \frac{d}{dt}(\alpha_t^* \omega)\Big|_{t=0}$$

where α is the flow of the vector field ξ, and we call $\mathscr{L}_\xi$ the **Lie derivative** again, applied to the differential form ω. We may rewrite this definition in terms of the value on vector fields $\xi_1, \dots, \xi_r$ as follows:

$$(\mathscr{L}_\xi \omega)(\xi_1, \dots, \xi_r) = \frac{d}{dt}\langle \omega \circ \alpha_t, \alpha_{t*}\xi_1 \times \cdots \times \alpha_{t*}\xi_r\rangle \Big|_{t=0}$$

Proposition 5.1. *Let ξ be a vector field and ω a differential form of degree $r \geq 1$. The Lie derivative $\mathscr{L}_\xi$ is a derivation, in the sense that*

$$\mathscr{L}_\xi\big(\omega(\xi_1, \dots, \xi_r)\big) = (\mathscr{L}_\xi \omega)(\xi_1, \dots, \xi_r) + \sum_{i=1}^{r} \omega(\xi_1, \dots, \mathscr{L}_\xi \xi_i, \dots, \xi_r)$$

where of course $\mathscr{L}_\xi \xi_i = [\xi, \xi_i]$.

If ξ, ξ_i, ω denote the local representations of the vector fields and the form respectively, then the Lie derivative $\mathscr{L}_\xi \omega$ has the local representation

$$\langle (\mathscr{L}_\xi \omega)(x), \xi_1(x) \times \cdots \times \xi_r(x)\rangle$$
$$= \langle \omega'(x)\xi(x), \xi_1(x) \times \cdots \times \xi_r(x)\rangle$$
$$+ \sum_{i=1}^{r} \langle \omega(x), \xi_1(x) \times \cdots \times \xi'(x)\xi_i(x) \times \cdots \times \xi_r(x)\rangle.$$

Proof. The proof is routine using the definitions. The first assertion is obvious by the definition of the pull back of a form. For the local expression we actually derive more, namely we derive a local expression for $\alpha_t^* \omega$ and $\frac{d}{dt}\alpha_t^* \omega$ which are characterized by their values at $(\xi_1, \dots, \xi_r)$. So we let

(1) $F(t) = \langle (\alpha_t^* \omega)(x), \xi_1(x) \times \cdots \times \xi_r(x)\rangle$
$$= \langle \omega\big(\alpha(t, x)\big), D_2\alpha(t, x)\xi_1(x) \times \cdots \times D_2\alpha(t, x)\xi_r(x)\rangle.$$

Then the Lie derivative $(\mathscr{L}_\xi \omega)(x)$ is precisely $F'(0)$, but we obtain also the local representation for $\frac{d}{dt}\alpha_t^* \omega$:

(2) $\quad F'(t) = \left\langle \dfrac{d}{dt}\alpha_t^*\omega(x),\ \xi_1(x) \times \cdots \times \xi_r(x) \right\rangle =$

(3) $\qquad\qquad \langle \omega'(\alpha(t,\,x))D_1\alpha(t,\,x),\ D_2\alpha(t,\,x)\xi_1(x) \times \cdots \times D_2\alpha(t,\,x)\xi_r(x)\rangle$

$\quad + \displaystyle\sum_{i=1}^{r} \langle \omega(\alpha(t,\,x)),\ D_2\alpha(t,\,x)\xi_1(x) \times \cdots \times D_1D_2\alpha(t,\,x)\xi_i(x) \times \cdots \times D_2\alpha(t,\,x)\xi_r(x)\rangle$

by the rule for the derivative of a product. Putting $t = 0$ and using the differential equation satisfied by $D_2\alpha(t, x)$, we get precisely the local expression as stated in the proposition. Remember the initial condition $D_2\alpha(0, x) = \mathrm{id}$.

From Proposition 5.1, we conclude that the Lie derivative gives an **R**-linear map

$$\mathscr{L}_\xi\colon\ \mathscr{A}^r(X) \to \mathscr{A}^r(X).$$

We may use expressions (1) and (3) in the above proof to derive a formula which holds even more generally for time-dependent vector fields.

Proposition 5.2. *Let ξ_t be a time-dependent vector field, α its flow, and let ω be a differential form. Then*

$$\frac{d}{dt}(\alpha_t^*\omega) = \alpha_t^*(\mathscr{L}_{\xi_t}\omega) \qquad or \qquad \frac{d}{dt}(\alpha_t^*\omega) = \alpha_t^*(\mathscr{L}_\xi\omega)$$

for a time-independent vector field.

Proof. Proposition 5.1 gives us a local expression for $(\mathscr{L}_{\xi_t}\omega)(y)$, replacing x by y because we shall now put $y = \alpha(t, x)$. On the other hand, from (1) in the proof of Proposition 5.1, we obtain

$$\alpha_t^*(\mathscr{L}_{\xi_t}\omega)(x) = \langle(\mathscr{L}_{\xi_t}\omega)(y),\ D_2\alpha(t,\,x)\xi_1(x) \times \cdots \times D_2\alpha(t,\,x)\xi_r(x)\rangle.$$

Substituting the local expression for $(\mathscr{L}_{\xi_t}\omega)(y)$, we get expression (3) from the proof of Proposition 5.1, thereby proving Proposition 5.2.

Proposition 5.3. *As a map on differential forms, the Lie derivative satisfies the following properties.*

LIE 1. $\mathscr{L}_\xi = d \circ C_\xi + C_\xi \circ d$, so $\mathscr{L}_\xi = C_\xi \circ d$ on functions.

LIE 2. $\mathscr{L}_\xi(\omega \wedge \psi) = \mathscr{L}_\xi\omega \wedge \psi + \omega \wedge \mathscr{L}_\xi\psi.$

LIE 3. $\mathscr{L}_\xi$ commutes with d and C_ξ.

LIE 4. $\mathscr{L}_{[\xi,\eta]} = \mathscr{L}_\xi \circ \mathscr{L}_\eta - \mathscr{L}_\eta \circ \mathscr{L}_\xi.$

LIE 5. $C_{[\xi,\eta]} = \mathscr{L}_\xi \circ C_\eta - C_\eta \circ \mathscr{L}_\xi.$

LIE 6. $\mathscr{L}_{f\xi}\omega = f\mathscr{L}_\xi\omega + df \wedge C_\xi\omega$ for all forms ω and functions f.

Proof. Let $\xi_1,\ldots,\xi_r$ be vector fields, and ω an r-form. Using the definition of the contraction and the local formula of Proposition 5.1, we find that $C_\xi \, d\omega$ is given locally by

$$\langle C_\xi \, d\omega(x), \xi_1(x) \times \cdots \times \xi_r(x)\rangle$$
$$= \langle \omega'(x)\xi(x), \xi_1(x) \times \cdots \times \xi_r(x)\rangle$$
$$+ \sum_{i=1}^{r}(-1)^i\langle \omega'(x)\xi_i(x), \xi(x) \times \xi_1(x) \times \cdots \times \widehat{\xi_i(x)} \times \cdots \xi_r(x)\rangle.$$

On the other hand, $dC_\xi\omega$ is given by

$$\langle dC_\xi\omega(x), \xi_1(x) \times \cdots \times \xi_r(x)\rangle$$
$$= \sum_{i=1}^{r}(-1)^{i+1}\langle (C_\xi\omega)'(x)\xi_i(x), \xi_1(x) \times \cdots \times \widehat{\xi_i(x)} \times \cdots \times \xi_r(x)\rangle.$$

To compute $(C_\xi\omega)'(x)$ is easy, going back to the definition of the derivative. At vectors $v_1,\ldots,v_{r-1}$, the form $C_\xi\omega(x)$ has the value

$$\langle\omega(x), \xi(x) \times v_1 \times \cdots \times v_{r-1}\rangle.$$

Differentiating this last expression with respect to x and evaluating at a vector h we get

$$\langle\omega'(x)h, \xi(x) \times v_1 \times \cdots \times v_{r-1}\rangle + \langle\omega(x), \xi'(x)h \times v_1 \times \cdots \times v_{r-1}\rangle.$$

Hence $\langle dC_\xi\omega(x), \xi_1(x) \times \cdots \times \xi_r(x)\rangle$ is equal to

$$\sum_{i=1}^{r}(-1)^{i+1}\langle\omega'(x)\xi_i(x), \xi(x) \times \xi_1(x) \times \cdots \times \widehat{\xi_i(x)} \times \cdots \times \xi_r(x)\rangle$$
$$+ \sum_{i=1}^{r}(-1)^{i+1}\langle\omega(x), \xi'(x)\xi_i(x) \times \xi_1(x) \times \cdots \times \widehat{\xi_i(x)} \times \cdots \times \xi_r(x)\rangle.$$

Shifting $\xi'(x)\xi_i(x)$ to the i-th place in the second sum contributes a sign of $(-1)^{i-1}$ which gives 1 when multiplied by $(-1)^{i+1}$. Adding the two local representations for $dC_\xi\omega$ and $C_\xi \, d\omega$, we find precisely the expression of Proposition 5.1, thus proving **LIE 1**.

As for **LIE 2**, it consists in using the derivation rule for d and C_ξ in Proposition 3.3, **EXD 1**, and **CON 3**. The corresponding rule for

$\mathscr{L}_\xi$ follows at once. (Terms will cancel just the right way.) The other properties are then clear.

V, §6. VECTOR FIELDS AND 1-FORMS UNDER SELF DUALITY

Let $\mathbf{E}$ be a vector space and let

$$(v, w) \mapsto \langle v, w \rangle$$

be a bilinear function of $\mathbf{E} \times \mathbf{E} \to \mathbf{R}$. We call such a function a **bilinear form**. This form induced a linear map

$$\lambda: \mathbf{E} \to \mathbf{E}^\vee$$

which to each $v \in \mathbf{E}$ associates the functional λ_v such that

$$\lambda_v(w) = \langle v, w \rangle.$$

We have a similar map on the other side. If both these mappings are linear isomorphisms of $\mathbf{E}$ and $\mathbf{E}^\vee$ then we say that the bilinear form is **non-singular**. Such a non-singular form exists, and we say that $\mathbf{E}$ is **self-dual** with respect to this form. For instance, a euclidean space is self-dual.

It suffices for a bilinear form to be non-singular that its kernels on the right and on the left be 0. (The kernels are the kernels of the associated maps λ as above.)

Let $\mathbf{E}$ be self dual with respect to the non-singular form $(v, w) \mapsto \langle v, w \rangle$, and let

$$\Omega: \mathbf{E} \times \mathbf{E} \to \mathbf{R}$$

be a continuous bilinear map. There exists a unique operator A such that

$$\Omega(v, w) = \langle Av, w \rangle$$

for all $v, w \in \mathbf{E}$. (An **operator** is a linear map $\mathbf{E} \to \mathbf{E}$ by definition.)

Remarks. *Suppose that the form* $(v, w) \mapsto \langle v, w \rangle$ *is symmetric*, i.e.

$$\langle v, w \rangle = \langle w, v \rangle$$

for all $v, w \in \mathbf{E}$. Then Ω is **symmetric** (resp. **alternating**) if and only if A is symmetric (resp. skew-symmetric). Recall that A symmetric (with respect to $\langle , \rangle)$ means that

$$\langle Av, w \rangle = \langle v, Aw \rangle \qquad \text{for all} \quad v, w \in \mathbf{E}.$$

That A is skew-symmetric means that $\langle Av, w \rangle = -\langle Aw, w \rangle$ for all $v, w \in \mathbf{E}$. For any operator $A \colon \mathbf{E} \to \mathbf{E}$ there is another operator ${}^{t}A$ (the transpose of A with respect to the non-singular form $\langle , \rangle$) such that for all $v, w \in \mathbf{E}$ we have

$$\langle Av, w \rangle = \langle v, {}^{t}Aw \rangle.$$

Thus A is symmetric (resp. skew-symmetric) if and only if ${}^{t}A = A$ (resp. ${}^{t}A = -A$).

The above remarks apply to any bilinear form Ω. For invertibility, we have the criterion:

The form Ω is non-singular if and only if the operator A representing the form with respect to $\langle , \rangle$ is invertible.

The easy verification is left to the reader. Of course, invertibility or non-singularity can be checked by verifying that the matrix representing the linear map with respect to bases has non-zero determinant. Similarly, the form is also represented by a matrix with respect to a choice of bases, and its being non-singular is equivalent to the matrix representing the form being invertible.

We recall that the set of invertible operators in $\mathrm{Laut}(\mathbf{E})$ is an open subset. Alternatively, the set of non-singular bilinear forms on $\mathbf{E}$ is an open subset of $L^{2}(\mathbf{E})$.

We may now globalize these notions to a vector bundle (and eventually especially to the tangent bundle) as follows.

Let X be a manifold, and $\pi \colon E \to X$ a vector bundle over X with fibers which are linearly isomorphic to $\mathbf{E}$, or as we shall also say, **modeled** on $\mathbf{E}$. Let Ω be a tensor field of type L^{2} on E, that is to say, a section of the bundle $L^{2}(E)$ $\big($or $L^{2}(\pi)\big)$, or as we shall also say, a **bilinear tensor field** on E. Then for each $x \in X$, we have a bilinear form Ω_{x} on E_{x}.

If Ω_{x} is non-singular for each $x \in X$ then we say that Ω is **non-singular**. If π is trivial, and we have a trivialisation $X \times \mathbf{E}$, then the local representation of Ω can be described by a morphism of X into the space of operators. If Ω is non-singular, then the image of this morphism is contained in the open set of invertible operators. (If Ω is a 2-form, this image is contained in the submanifold of skew-symmetric operators.) For example, in a chart U, we can represent Ω over U by a morphism

$$A \colon U \to L(\mathbf{E}, \mathbf{E}) \qquad \text{such that} \qquad \Omega_{x}(v, w) = \langle A_{x}v, w \rangle$$

for all $v, w \in \mathbf{E}$. Here we wrote A_{x} instead of $A(x)$ to simplify the typography.

A non-singular Ω as above can be used to establish a linear isomorphism

$$\Gamma(E) \to \Gamma L^1(E), \qquad \text{also denoted by} \quad \Gamma L(E) \text{ or } \Gamma E^{\vee},$$

between the (infinite dimensional) **R**-vector spaces of sections $\Gamma(E)$ of E and the 1-forms on E in the following manner. Let ξ be a section of E. For each $x \in X$ we define a continuous linear map

$$(\Omega \circ \xi)_x \colon E_x \to \mathbf{R}$$

by the formula

$$(\Omega \circ \xi)_x(w) = \Omega_x(\xi(x), w).$$

Looking at local trivialisations of π, we see at once that $\Omega \circ \xi$ is a 1-form on E.

Conversely, let ω be a given 1-form on E. For each $x \in X$, ω_x is therefore a 1-form on E_x and since Ω is non-singular, there exists a unique element $\xi(x)$ of E_x such that

$$\Omega_x(\xi(x), w) = \omega_x(w)$$

for all $w \in E_x$. In this fashion, we obtain a mapping ξ of X into E and we contend that ξ is a morphism (and therefore a section).

To prove our contention we can look at the local representations. We use Ω and ω to denote these. They are represented over a suitable open set U by two morphisms

$$A \colon U \to \operatorname{Aut}(\mathbf{E}) \qquad \text{and} \qquad \eta \colon U \to \mathbf{E}$$

such that

$$\Omega_x(v, w) = \langle A_x v, w \rangle \qquad \text{and} \qquad \omega_x(w) = \langle \eta(x), w \rangle.$$

From this we see that

$$\xi(x) = A_x^{-1} \eta(x),$$

from which it is clear that ξ is a morphism. We may summarize our discussion as follows.

Proposition 6.1. *Let X be a manifold and $\pi \colon E \to X$ a vector bundle over X modeled on $\mathbf{E}$. Let Ω be a non-singular bilinear tensor field on E. Then Ω induces an isomorphism of $\operatorname{Fu}(X)$-modules*

$$\Gamma E \to \Gamma E^{\vee}.$$

A section ξ corresponds to a 1-form ω if and only if $\Omega \circ \xi = \omega$.

In many applications, one takes the differential form to be df for some function f. The vector field corresponding to df is then called the **gradient of f with respect to Ω**.

Remark. There is no universally accepted notation to denote the correspondence between a 1-form and a vector field under Ω as above. Some authors use sharps and flats, which have two disadvantages. First, they do not provide a symbol for the mapping, and second they do not contain the Ω in the notation. I would propose the check sign $\bigvee_\Omega$ to denote either isomorphism

$$\bigvee_\Omega: \Gamma L(E) \to \Gamma E \qquad \text{denoted on elements by} \qquad \omega \mapsto \bigvee_\Omega \omega = \omega^\vee = \xi_\omega$$

and also

$$\bigvee_\Omega: \Gamma E \to \Gamma L(E) \qquad \text{denoted on elements by} \qquad \xi \mapsto \bigvee_\Omega \xi = \xi^\vee = \omega_\xi.$$

If Ω is fixed throughout a discussion and need not be referred to, then it is useful to write $\xi^\vee$ or $\lambda^\vee$ in some formulas. We have $\bigvee_\Omega \circ \bigvee_\Omega = \text{id}$. Instead of the sharp and flat superscript, I prefer the single $^\vee$ sign.

Many important applications of the above duality occur when Ω is a non-singular symmetric bilinear tensor field on the tangent bundle TX. Such a tensor field is then usually denoted by g. If ξ, η are vector fields, we may then define their scalar product to be the function

$$\langle \xi, \eta \rangle_g = g(\xi, \eta).$$

On the other hand, by the duality of Proposition 6.1, if i.e. ω, λ are 1-forms, i.e. sections of the dual bundle $T^\vee X$, then $\omega^\vee$ and $\lambda^\vee$ are vector fields, and we define the scalar product of the 1-forms to be

$$\langle \omega, \lambda \rangle_g = \langle \omega^\vee, \lambda^\vee \rangle_g.$$

This duality is especially important for Riemannian metrics, as in Chapter X.

The rest of this section will not be used in the book.

In Proposition 6.1, we dealt with a quite general non-singular bilinear tensor field on E. We now specialize to the case when $E = TX$ is the tangent bundle of X, and Ω is a 2-form, i.e. Ω is alternating. A pair (X, Ω) consisting of a manifold and a non-singular closed 2-form is called a **symplectic manifold**. (Recall that **closed** means $d\Omega = 0$.)

We denote by ξ, η vector fields over X, and by f, h functions on X, so that df, dh are 1-forms. We let ξ_{df} be the vector field on X which corresponds to df under the 2-form Ω, according to Proposition 6.1. Vector fields on X which are of type ξ_{df} are called **Hamiltonian** (with respect to the 2-form). More generally, we denote by ξ_ω the vector field corresponding to a 1-form ω. By definition we have the formula

Ω 1. $\Omega \circ \xi_\omega = \omega$ so in particular $\Omega \circ \xi_{df} = df$.

In Chapter VII, §7 we shall consider a particularly important example, when the base manifold is the cotangent bundle; the function is the **kinetic energy**

$$K(v) = \tfrac{1}{2} \langle v, v \rangle_g$$

with respect to the scalar product g of a Riemannian or pseudo Riemannian metric, and the 2-form Ω arises canonically from the pseudo Riemannian metric.

In general, by **LIE 1** of Proposition 5.3 formula Ω 1, and the fact that $d\Omega = 0$, we find for any 1-form ω that:

Ω 2. $\mathscr{L}_{\xi_\omega} \Omega = d\omega.$

The next proposition reinterprets this formula in terms of the flow when $d\omega = 0$.

Proposition 6.2. *Let ω be such that $d\omega = 0$. Let α be the flow of ξ_ω. Then $\alpha_t^* \Omega = \Omega$ for all t (in the domain of the flow).*

Proof. By Proposition 5.2,

$$\frac{d}{dt} \alpha_t^* \Omega = \alpha_t^* \mathscr{L}_{\xi_\omega} \Omega = 0 \quad \text{by } \Omega \text{ 2.}$$

Hence $\alpha_t^* \Omega$ is constant, equal to $\alpha_0^* \Omega = \Omega$, as was to be shown.

A special case of Proposition 6.2 in Hamiltonian mechanics is when $\omega = dh$ for some function h. Next by **LIE 5**, we obtain for any vector fields ξ, η:

$$\mathscr{L}_\xi (\Omega \circ \eta) = (\mathscr{L}_\xi \Omega) \circ \eta + \Omega \circ [\xi, \eta].$$

In particular, since $ddf = 0$, we get

Ω 3. $\mathscr{L}_{\xi_{df}} (\Omega \circ \xi_{dh}) = \Omega \circ [\xi_{df}, \xi_{dh}].$

One defines the **Poisson bracket** between two functions f, h to be

$$\{f, h\} = \xi_{df} \cdot h.$$

Then the preceding formula may be rewritten in the form

Ω 4. $[\xi_{df}, \xi_{dh}] = \xi_{d\{f,h\}}.$

It follows immediately from the definitions and the antisymmetry of the ordinary bracket between vector fields that the Poisson bracket is also antisymmetric, namely

$$\{f, h\} = -\{h, f\}.$$

In particular, we find that

$$\xi_{df} \cdot f = 0.$$

In the case of the cotangent bundle with a symplectic 2-form as in the next section, physicists think of f as an energy function, and interpret this formula as a law of conservation of energy. The formula expresses the property that f is constant on the integral curves of the vector field ξ_{df}. This property follows at once from the definition of the Lie derivative of a function. Furthermore:

Proposition 6.3. *If* $\xi_{df} \cdot h = 0$ *then* $\xi_{dh} \cdot f = 0$.

This is immediate from the antisymmetry of the Poisson bracket. It is interpreted as conservation of momentum in the physical theory of Hamiltonian mechanics, when one deals with the canonical 2-form on the cotangent bundle, to be defined in the next section.

V, §7. THE CANONICAL 2-FORM

Consider the functor $\mathbf{E} \mapsto L(\mathbf{E})$ (linear forms). If $E \to X$ is a vector bundle, then $L(E)$ will be called the **dual bundle**, and will be denoted by $E^\vee$. For each $x \in X$, the fiber of the dual bundle is simply $L(E_x)$.

If $E = T(X)$ is the tangent bundle, then its dual is denoted by $T^\vee(X)$ and is called the **cotangent bundle**. Its elements are called **cotangent vectors**. The fiber of $T^\vee(X)$ over a point x of X is denoted by $T_x^\vee(X)$. For each $x \in X$ we have a pairing

$$T_x^\vee \times T_x \to \mathbf{R}$$

given by

$$\langle \lambda, u \rangle = \lambda(u)$$

for $\lambda \in T_x^\vee$ and $u \in T_x$ (it is the value of the linear form λ at u).

We shall now describe how to construct a canonical 1-form on the cotangent bundle $T^{\vee}(X)$. For each $\lambda \in T^{\vee}(X)$ we must define a 1-form on $T_{\lambda}(T^{\vee}(X))$.

Let $\pi: T^{\vee}(X) \to X$ be the canonical projection. Then the induced tangent map

$$T\pi = \pi_*: T(T^{\vee}(X)) \to T(X)$$

can be applied to an element z of $T_v(T^{\vee}(X))$ and one sees at once that $\pi_* z$ lies in $T_x(X)$ if λ lies in $T_x^{\vee}(X)$. Thus we can take the pairing

$$\langle \lambda, \pi_* z \rangle = \theta_{\lambda}(z)$$

to define a map (which is obviously continuous linear):

$$\theta_{\lambda}: T_{\lambda}(T^{\vee}(X)) \to \mathbf{R}.$$

Proposition 7.1. *This map defines a 1-form on $T^{\vee}(X)$. Let $X = U$ be open in $\mathbf{E}$ and*

$$T^{\vee}(U) = U \times \mathbf{E}^{\vee}, \quad T(T^{\vee}(U)) = (U \times \mathbf{E}^{\vee}) \times (\mathbf{E} \times \mathbf{E}^{\vee}).$$

If $(x, \lambda) \in U \times \mathbf{E}^{\vee}$ and $(u, \omega) \in \mathbf{E} \times \mathbf{E}^{\vee}$, then the local representation $\theta_{(x,\lambda)}$ is given by

$$\langle \theta_{(x,\lambda)}, (u, \omega) \rangle = \lambda(u).$$

Proof. We observe that the projection $\pi: U \times \mathbf{E}^{\vee} \to U$ is linear, and hence that its derivative at each point is constant, equal to the projection on the first factor. Our formula is then an immediate consequence of the definition. The local formula shows that θ is in fact a 1-form locally, and therefore globally since it has an invariant description.

Our 1-form is called the **canonical 1-form on the cotangent bundle**. We define the **canonical 2-form** Ω on the cotangent bundle $T^{\vee}X$ to be

$$\Omega = -d\theta.$$

The next proposition gives a local description of Ω.

Proposition 7.2. *Let U be open in $\mathbf{E}$, and let Ω be the local representation of the canonical 2-form on $T^{\vee}U = U \times \mathbf{E}^{\vee}$. Let $(x, \lambda) \in U \times \mathbf{E}^{\vee}$. Let (u_1, ω_1) and (u_2, ω_2) be elements of $\mathbf{E} \times \mathbf{E}^{\vee}$. Then*

$$\langle \Omega_{(x,\lambda)}, (u_1, \omega_1) \times (u_2, \omega_2) \rangle = \langle u_1, \omega_2 \rangle - \langle u_2, \omega_1 \rangle$$

$$= \omega_2(u_1) - \omega_1(u_2).$$

Proof. We observe that θ is linear, and thus that θ' is constant. We then apply the local formula for the exterior derivative, given in Proposition 3.2. Our assertion becomes obvious.

The canonical 2-form plays a fundamental role in Lagrangian and Hamiltonian mechanics, cf. [AbM 78], Chapter 3, §3. I have taken the sign of the canonical 2-form both so that its value is a 2×2 determinant, and so that it fits with, for instance, [LoS 68] and [AbM 78]. We observe that Ω is closed, that is $d\Omega = 0$, because $\Omega = -d\theta$. Thus $(T^{\vee} X, \Omega)$ is a symplectic manifold, to which the properties listed at the end of the last section apply.

In particular, let ξ be a vector field on X. Then to ξ is **associated a function** called the **momentum function**

$$f_{\xi} \colon T^{\vee} X \to \mathbf{R} \qquad \text{such that} \qquad f_{\xi}(\lambda_x) = \lambda_x\big(\xi(x)\big)$$

for $\lambda_x \in T_x^{\vee} X$. Then df_{ξ} is a 1-form on $T^{\vee} X$. Classical Hamiltonian mechanics then applies Propositions 6.2 and 6.3 to this situation. We refer the interested reader to [LoS 68] and [AbM 78] for further information on this topic. For an important theorem of Marsden–Weinstein [MaW 74] and applications to vector bundles, see [Ko 87].

V, §8. DARBOUX'S THEOREM

If $\mathbf{E} = \mathbf{R}^n$ then the usual scalar product establishes the self-duality of $\mathbf{R}^n$. This self-duality arises from other forms, and in this section we are especially interested in the self-duality arising from alternating forms. If $\mathbf{E}$ is finite dimensional and ω is an element of $L_a^2(\mathbf{E})$, that is an alternating 2-form, which is non-singular, then one sees easily that the dimension of $\mathbf{E}$ is even.

Example. An example of such a form on $\mathbf{R}^{2n}$ is the following. Let

$$v = (v_1, \ldots, v_n, v_1', \ldots, v_n'),$$
$$w = (w_1, \ldots, w_n, w_1', \ldots, w_n'),$$

be elements of $\mathbf{R}^{2n}$, with components v_i, v_i', w_i, w_i'. Letting

$$\omega(v, w) = \sum_{i=1}^{n} (v_i w_i' - v_i' w_i)$$

defines a non-singular 2-form ω on $\mathbf{R}^{2n}$. It is an exercise of linear algebra to prove that any non-singular 2-form on $\mathbf{R}^{2n}$ is linearly isomorphic to this particular one in the following sense. If

$$f\colon E \to F$$

is a linear isomorphism between two finite dimensional spaces, then it induces an isomorphism

$$f^*\colon L_a^2(F) \to L_a^2(E).$$

We call forms ω on E and ψ on F **linearly isomorphic** if there exists a linear isomorphism f such that $f^*\psi = \omega$. Thus up to a linear isomorphism, there is only one non-singular 2-form on $\mathbf{R}^{2n}$. (For a proof, cf. for instance my book *Algebra*.)

We are interested in the same question on a manifold locally. Let U be open in the Banach space $\mathbf{E}$ and let $x_0 \in U$. A 2-form

$$\omega\colon U \to L_a^2(\mathbf{E})$$

is said to be **non-singular** if each form $\omega(x)$ is non-singular. If ξ is a vector field on U, then $\omega \circ \xi$ is a 1-form, whose value at (x, ω) is given

$$(\omega \circ \xi)(x)(w) = \omega(x)(\xi(x), w).$$

As a special case of Proposition 6.1, we have:

Let ω be a non-singular 2-form on an open set U in $\mathbf{E}$. The association

$$\xi \mapsto \omega \circ \xi$$

is a linear isomorphism between the space of vector fields on U and the space of 1-forms on U.

Let

$$\omega\colon U \to L_a^2(U)$$

be a 2-form on an open set U in $\mathbf{E}$. If there exists a local isomorphism f at a point $x_0 \in U$, say

$$f\colon U_1 \to V_1,$$

and a 2-form ψ on V_1 such that $f^*\psi = \omega$ (or more accurately, ω restricted to U_1), then we say that ω is **locally isomorphic** to ψ at x_0. Observe that in the case of an isomorphism we can take a direct image of forms, and we shall also write

$$f_*\omega = \psi$$

instead of $\omega = f^*\psi$. In other words, $f_* = (f^{-1})^*$.

Example. On $\mathbf{R}^{2n}$ we have the constant form of the previous example. In terms of local coordinates $(x_1, \ldots, x_n, y_1, \ldots, y_n)$, this form has the local expression

$$\omega(x, y) = \sum_{i=1}^{n} dx_i \wedge dy_i.$$

This 2-form will be called the **standard 2-form** on $\mathbf{R}^{2n}$.

The Darboux theorem states that any non-singular closed 2-form in $\mathbf{R}^{2n}$ is locally isomorphic to the standard form, that is that in a suitable chart at a point, it has the standard expression of the above example. A technique to show that certain forms are isomorphic was used by Moser [Mo 65], who pointed out that his arguments also prove the classical Darboux theorem.

Theorem 8.1 (Darboux Theorem). *Let*

$$\omega: \ U \to L_a^2(\mathbf{E})$$

be a non-singular closed 2-form on an open set of $\mathbf{E}$, *and let* $x_0 \in U$. *Then* ω *is locally isomorphic at* x_0 *to the constant form* $\omega(x_0)$.

Proof. Let $\omega_0 = \omega(x_0)$, and let

$$\omega_t = \omega_0 + t(\omega - \omega_0), \qquad 0 \leq t \leq 1.$$

We wish to find a time-dependent vector field ξ_t locally at 0 such that if α denotes its flow, then

$$\alpha_t^* \omega_t = \omega_0.$$

Then the local isomorphism α_1 satisfies the requirements of the theorem. By the Poincaré lemma, there exists a 1-form θ locally at 0 such that

$$\omega - \omega_0 = d\theta,$$

and without loss of generality, we may assume that $\theta(x_0) = 0$. We contend that the time-dependent vector field ξ_t, such that

$$\omega_t \circ \xi_t = -\theta,$$

has the desired property. Let α be its flow. If we shrink the domain of the vector field near x_0 sufficiently, and use the fact that $\theta(x_0) = 0$, then we can use the local existence theorem (Proposition 1.1 of Chapter IV) to see that the flow can be integrated at least to $t = 1$ for all points x in this small domain. We shall now verify that

$$\frac{d}{dt}(\alpha_t^* \omega_t) = 0.$$

This will prove that $\alpha_t^* \omega_t$ is constant. Since we have $\alpha_0^* \omega_0 = \omega_0$ because

$$\alpha(0, x) = x \qquad \text{and} \qquad D_2 \alpha(0, x) = \text{id},$$

it will conclude the proof of the theorem.

We compute locally. We use the local formula of Proposition 5.2, and formula **LIE 1**, which reduces to

$$\mathscr{L}_{\xi_t} \omega_t = d(\omega_t \circ \xi_t),$$

because $d\omega_t = 0$. We find

$$\frac{d}{dt}(\alpha_t^* \omega_t) = \alpha_t^* \left(\frac{d}{dt}\omega_t\right) + \alpha_t^* (\mathscr{L}_{\xi_t} \omega_t)$$

$$= \alpha_t^* \left(\frac{d}{dt}\omega_t + d(\omega_t \circ \xi_t)\right)$$

$$= \alpha_t^* (\omega - \omega_0 - d\theta)$$

$$= 0.$$

This proves Darboux's theorem.

Remark 1. For the analogous uniqueness statement in the case of a non-singular symmetric form, see the Morse–Palais lemma of Chapter VII, §5.

Remark 2. The proof of the Poincaré lemma can also be cast in the above style. For instance, let $\phi_t(x) = tx$ be a retraction of a star shaped open set around 0. Let ξ_t be the vector field whose flow is ϕ_t, and let ω be a closed form. Then

$$\frac{d}{dt}\phi_t^* \omega = \phi_t^* \mathscr{L}_{\xi_t} \omega = \phi_t^* \, dC_{\xi_t}\omega = d\phi_t^* C_{\xi_t} \omega.$$

Since $\phi_0^* \omega = 0$ and ϕ_1 is the identity, we see that

$$\omega = \phi_1^* \omega - \phi_0^* \omega = \int_0^1 \frac{d}{dt}\phi_t^* \omega \, dt = d \int_0^1 \phi_t^* C_{\xi_t} \omega \, dt$$

is exact, thus concluding a proof of Poincaré's theorem.

CHAPTER VI

The Theorem of Frobenius

Having acquired the language of vector fields, we return to differential equations and give a generalization of the local existence theorem known as the Frobenius theorem, whose proof will be reduced to the standard case discussed in Chapter IV. We state the theorem in §1. Readers should note that one needs only to know the definition of the bracket of two vector fields in order to understand the proof. It is convenient to insert also a formulation in terms of differential forms, for which the reader needs to know the local definition of the exterior derivative. However, the condition involving differential forms is proved to be equivalent to the vector field condition at the very beginning, and does not reappear explicitly afterwards.

We shall follow essentially the proof given by Dieudonné in his *Foundations of Modern Analysis*, allowing for the fact that we use freely the geometric language of vector bundles, which is easier to grasp.

It is convenient to recall in §2 the statements concerning the existence theorems for differential equations depending on parameters. The proof of the Frobenius theorem proper is given in §3. An important application to Lie groups is given in §5, after formulating the theorem globally.

The present chapter will not be used in the rest of this book.

VI, §1. STATEMENT OF THE THEOREM

Let X be a manifold of class C^p $(p \geq 2)$. A subbundle E of its tangent bundle will also be called a **tangent subbundle** over X. We contend that the following two conditions concerning such a subbundle are equivalent.

FR 1. *For each point $z \in X$ and vector fields ξ, η at z (i.e. defined on an open neighborhood of z) which lie in E (i.e. such that the image of each point of X under ξ, η lies in E), the bracket $[\xi, \eta]$ also lies in E.*

FR 2. *For each point $z \in X$ and differential form ω of degree 1 at z which vanishes on E, the form $d\omega$ vanishes on $\xi \times \eta$ whenever ξ, η are two vector fields at z which lie in E.*

The equivalence is essentially a triviality. Indeed, assume **FR 1**. Let ω vanish to E. Then

$$\langle d\omega, \xi \times \eta \rangle = -\langle \omega, [\xi, \eta] \rangle - \eta\langle \omega, \xi \rangle + \xi\langle \omega, \eta \rangle.$$

By assumption the right-hand side is 0 when evaluated at z. Conversely, assume **FR 2**. Let ξ, η be two vector fields at z lying in E. If $[\xi, \eta](z)$ is not in E, then we see immediately from a local product representation that there exists a differential form ω of degree 1 defined on a neighborhood of z which is 0 on E_z and non-zero on $[\xi, \eta](z)$, thereby contradicting the above formula.

We shall now give a third condition equivalent to the above two, and actually, we shall not refer to **FR 2** any more. We remark merely that, it is easy to prove that when a differential form ω satisfies condition **FR 2**, then $d\omega$ can be expressed locally in a neighborhood of each point as a finite sum

$$d\omega = \sum \gamma_i \wedge \omega_i$$

where γ_i and ω_i are of degree 1 and each ω_i vanishes on E. We leave this as an exercise to the reader.

Let E be a tangent subbundle over X. We shall say that E is **integrable** at a point x_0 if there exists a submanifold Y of X containing x_0 such that the tangent map of the inclusion

$$j\colon Y \to X$$

induces a VB-isomorphism of TY with the subbundle E restricted to Y. Equivalently, we could say that for each point $y \in Y$, the tangent map

$$T_y j\colon T_y Y \to T_y X$$

induces a linear isomorphism of $T_y Y$ on E_y. Note that our condition defining integrability is local at x_0. We say that E is **integrable** if it is integrable at every point.

Using the functoriality of vector fields, and their relations under tangent maps and the bracket product, we see at once that if E is integrable, then it satisfies **FR 1**. Indeed, locally vector fields having their values in E are related to vector fields over Y under the inclusion mapping.

Frobenius' theorem asserts the converse.

Theorem 1.1. *Let X be a manifold of class C^p $(p \geq 2)$ and let E be a tangent subbundle over X. Then E is integrable if and only if E satisfies condition* **FR 1**.

The proof of Frobenius' theorem will be carried out by analyzing the situation locally and reducing it to the standard theorem for ordinary differential equations. Thus we now analyze the condition **FR 1** in terms of its local representation.

Suppose that we work locally, over a product $U \times V$ of open subsets of vector spaces **E** and **F**. Then the tangent bundle $T(U \times V)$ can be written in a natural way as a direct sum. Indeed, for each point (x, y) in $U \times V$ we have

$$T_{(x, y)}(U \times V) = T_x(U) \times T_y(V).$$

One sees at once that the collection of fibers $T_x(U) \times 0$ (contained in $T_x(U) \times T_y(V)$) forms a subbundle which will be denoted by $T_1(U \times V)$ and will be called the **first factor** of the tangent bundle. One could define $T_2(U \times V)$ similarly, and

$$T(U \times V) = T_1(U \times V) \oplus T_2(U \times V).$$

A subbundle E of $T(X)$ is integrable at a point $z \in X$ if and only if there exists an open neighborhood W of z and an isomorphism

$$\varphi \colon U \times V \to W$$

of a product onto W such that the composition of maps

$$T_1(U \times V) \xrightarrow{\text{inc.}} T(U \times V) \xrightarrow{T\varphi} T(W)$$

induces a VB-isomorphism of $T_1(U \times V)$ onto $E|W$ (over φ). Denoting by φ_y the map of U into W given by $\varphi_y(x) = \varphi(x, y)$, we can also express the integrability condition by saying that $T_x\varphi_y$ should induce a linear isomorphism of **E** onto $E_{\varphi(x, y)}$ for all (x, y) in $U \times V$. We note that in terms of our local product structure, $T_x\varphi_y$ is nothing but the partial derivative $D_1\varphi(x, y)$.

Given a subbundle of $T(X)$, and a point in the base space X, we know from the definition of a subbundle in terms of a local product decomposition that we can find a product decomposition of an open neigh-borhood of this point, say $U \times V$, such that the point has coordinates (x_0, y_0) and such that the subbundle can be written in the form of an exact sequence

$$0 \to U \times V \times \mathbf{E} \xrightarrow{\tilde{f}} U \times V \times \mathbf{E} \times \mathbf{F}$$

with the map

$$f(x_0, y_0): \mathbf{E} \to \mathbf{E} \times \mathbf{F}$$

equal to the canonical embedding of $\mathbf{E}$ on $\mathbf{E} \times 0$. For a point (x, y) in $U \times V$ the map $f(x, y)$ has two components $f_1(x, y)$ and $f_2(x, y)$ into $\mathbf{E}$ and $\mathbf{F}$ respectively. Taking a suitable VB-automorphism of $U \times V \times \mathbf{E}$ if necessary, we may assume without loss of generality that $f_1(x, y)$ is the identity. We now write $f(x, y) = f_2(x, y)$. Then

$$f: U \times V \to L(\mathbf{E}, \mathbf{F})$$

is a morphism (of class C^{p-1}) which describes our subbundle completely. We shall interpret condition **FR 1** in terms of the present situation. If

$$\xi: U \times V \to \mathbf{E} \times \mathbf{F}$$

is the local representation of a vector field over $U \times V$, we let ξ_1 and ξ_2 be its projections on $\mathbf{E}$ and $\mathbf{F}$ respectively. Then ξ lies in the image of $\tilde{f}$ if and only if

$$\xi_2(x, y) = f(x, y)\xi_1(x, y)$$

for all (x, y) in $U \times V$, or in other words, if and only if ξ is of the form

$$\xi(x, y) = \big(\xi_1(x, y), f(x, y)\xi_1(x, y)\big)$$

for some morphism (of class C^{p-1})

$$\xi_1: U \times V \to \mathbf{E}.$$

We shall also write the above condition symbolically, namely

(1) $$\xi = (\xi_1, f \cdot \xi_1).$$

If ξ, η are the local representations of vector fields over $U \times V$, then the reader will verify at once from the local definition of the bracket (Proposition 1.3 of Chapter V) that $[\xi, \eta]$ lies in the image of $\tilde{f}$ if and only if

$$Df(x, y) \cdot \xi(x, y) \cdot \eta_1(x, y) = Df(x, y) \cdot \eta(x, y) \cdot \xi_1(x, y)$$

or symbolically,

(2) $$Df \cdot \xi \cdot \eta_1 = Df \cdot \eta \cdot \xi_1.$$

We have now expressed all the hypotheses of Theorem 1.1 in terms of local data, and the heart of the proof will consist in proving the following result.

Theorem 1.2. *Let U, V be open subsets of vector spaces $\mathbf{E}$, $\mathbf{F}$ respectively. Let*

$$f: \ U \times V \rightarrow L(\mathbf{E}, \mathbf{F})$$

be a C^r-morphism $(r \geqq 1)$. Assume that if

$$\xi_1, \eta_1: \ U \times V \rightarrow \mathbf{E}$$

are two morphisms, and if we let

$$\xi = (\xi_1, f \cdot \xi_1) \qquad and \qquad \eta = (\eta_1, f \cdot \eta_1)$$

then relation (2) above is satisfied. Let (x_0, y_0) be a point of $U \times V$. Then there exists open neighborhoods U_0, V_0 of x_0, y_0 respectively, contained in U, V, and a unique morphism $\alpha: \ U_0 \times V_0 \rightarrow V$ such that

$$D_1\alpha(x, y) = f\big(x, \alpha(x, y)\big)$$

and $\alpha(x_0, y) = y$ for all (x, y) in $U_0 \times V_0$.

We shall prove Theorem 1.2 in §3. We now indicate how Theorem 1.1 follows from it. We denote by α_y the map $\alpha_y(x) = \alpha(x, y)$, viewed as a map of U_0 into V. Then our differential equation can be written

$$D\alpha_y(x) = f\big(x, \alpha_y(x)\big).$$

We let

$$\varphi: \ U_0 \times V_0 \rightarrow U \times V$$

be the map $\varphi(x, y) = \big(x, \alpha_y(x)\big)$. It is obvious that $D\varphi(x_0, y_0)$ is a toplinear isomorphism, so that φ is a local isomorphism at (x_0, y_0). Furthermore, for $(u, v) \in \mathbf{E} \times \mathbf{F}$ we have

$$D_1\varphi(x, y) \cdot (u, v) = \big(u, D\alpha_y(x) \cdot u\big) = \big(u, f(x, \alpha_y(x)) \cdot u\big)$$

which shows that our subbundle is integrable.

VI, §2. DIFFERENTIAL EQUATIONS DEPENDING ON A PARAMETER

Proposition 2.1. *Let* U, V *be open sets in vector spaces* $\mathbf{E}$, $\mathbf{F}$ *respectively. Let* J *be an open interval of* $\mathbf{R}$ *containing* 0, *and let*

$$g: J \times U \times V \to \mathbf{F}$$

be a morphism of class C^r ($r \geqq 1$). *Let* (x_0, y_0) *be a point in* $U \times V$. *Then there exists open balls* J_0, U_0, V_0 *centered at* 0, x_0, y_0 *and contained* J, U, V *respectively, and a unique morphism of class* C^r

$$\beta: J_0 \times U_0 \times V_0 \to V$$

such that $\beta(0, x, y) = y$ *and*

$$D_1\beta(t, x, y) = g\big(t, x, \beta(t, x, y)\big)$$

for all $(t, x, y) \in J_0 \times U_0 \times V_0$.

Proof. This follows from the existence and uniqueness of local flows, by considering the ordinary vector field on $U \times V$

$$G: J \times U \times V \to \mathbf{E} \times \mathbf{F}$$

given by $G(t, x, y) = \big(0, g(t, x, y)\big)$. If $B(t, x, y)$ is the local flow for G, then we let $\beta(t, x, y)$ be the projection on the second factor of $B(t, x, y)$. The reader will verify at once that β satisfies the desired conditions. The uniqueness is clear.

Let us keep the initial condition y fixed, and write

$$\beta(t, x) = \beta(t, x, y).$$

From Chapter IV, §1, we obtain also the differential equation satisfied by β in its second variable:

Proposition 2.2. *Let notation be as in Proposition 2.1, and with* y *fixed, let* $\beta(t, x) = \beta(t, x, y)$. *Then* $D_2\beta(t, x)$ *satisfies the differential equation*

$$D_1 D_2\beta(t, x) \cdot v = D_2 g\big(t, x, \beta(t, x)\big) \cdot v + D_3 g\big(t, x, \beta(t, x)\big) \cdot D_2\beta(t, x) \cdot v,$$

for every $v \in \mathbf{E}$.

Proof. Here again, we consider the vector field as in the proof of Proposition 2.1, and apply the formula for the differential equation satisfied by $D_2\beta$ as in Chapter IV, §1.

VI, §3. PROOF OF THE THEOREM

In the application of Proposition 2.1 to the proof of Theorem 1.2, we take our morphism g to be

$$g(t, z, y) = f(x_0 + tz, y) \cdot z$$

with z in a small ball $\mathbf{E}_0$ around the origin in $\mathbf{E}$, and y in V. It is convenient to make a translation, and without loss of generality we can assume that $x_0 = 0$ and $y_0 = 0$. From Proposition 2.1 we then obtain

$$\beta\colon J_0 \times \mathbf{E}_0 \times V_0 \to V$$

with initial condition $\beta(0, z, y) = y$ for all $z \in \mathbf{E}_0$, satisfying the differential equation

$$D_1\beta(t, z, y) = f(tz, \beta(t, z, y)) \cdot z.$$

Making a change of variables of type $t = as$ and $z = a^{-1}x$ for a small positive number a, we see at once that we may assume that J_0 contains 1, provided we take $\mathbf{E}_0$ sufficiently small. As we shall keep y fixed from now on, we omit it from the notation, and write $\beta(t, z)$ instead of $\beta(t, z, y)$. Then our differential equation is

$$(3) \qquad\qquad D_1\beta(t, z) = f(tz, \beta(t, z)) \cdot z.$$

We observe that if we knew the existence of α in the statement of Theorem 1.2, then letting $\beta(t, z) = \alpha(x_0 + tz)$ would yield a solution of our differential equation. Thus the uniqueness of α follows. To prove its existence, we start with β and contend that the map

$$\alpha(x) = \beta(1, x)$$

has the required properties for small $|x|$. To prove our contention it will suffice to prove that

$$(4) \qquad\qquad D_2\beta(t, z) = tf(tz, \beta(t, z))$$

because if that relation holds, then

$$D\alpha(x) = D_2\beta(1, x) = f(x, \beta(1, x)) = f(x, \alpha(x))$$

which is precisely what we want.

From Proposition 2.2, we obtain for any vector $v \in \mathbf{E}$,

$$D_1 D_2 \beta(t, z) \cdot v = t D_1 f\bigl(tz, \beta(t, z)\bigr) \cdot v \cdot z$$
$$+ D_2 f\bigl(tz, \beta(t, z)\bigr) \cdot D_2 \beta(t, z) \cdot v \cdot z + f\bigl(tz, \beta(t, z)\bigr) \cdot v.$$

We now let $k(t) = D_2 \beta(t, z) \cdot v - t f\bigl(tz, \beta(t, z)\bigr) \cdot v$. Then one sees at once that $k(0) = 0$ and we contend that

$$(5) \qquad\qquad Dk(t) = D_2 f\bigl(tz, \beta(t, z)\bigr) \cdot k(t) \cdot z.$$

We use the main hypothesis of our theorem, namely relation (2), in which we take ξ_1 and η_1 to be the fields v and z respectively. We compute Df using the formula for the partial derivatives, and apply it to this special case. Then (5) follows immediately. It is a linear differential equation satisfied by $k(t)$, and by Corollary 1.7 of Chapter IV, we know that the solution 0 is the unique solution. Thus $k(t) = 0$ and relation (4) is proved. The theorem also.

VI, §4. THE GLOBAL FORMULATION

Let X be a manifold. Let F be a tangent subbundle. By an **integral manifold** for F, we shall mean an injective immersion

$$f: Y \to X$$

such that at every point $y \in Y$, the tangent map

$$T_y f: T_y Y \to T_{f(y)} X$$

induces a linear isomorphism of $T_y Y$ on the subspace $F_{f(y)}$ of $T_{f(y)} X$. Thus Tf induces locally an isomorphism of the tangent bundle of Y with the bundle F over $f(Y)$.

Observe that the image $f(Y)$ itself may not be a submanifold of X. For instance, if F has dimension 1 (i.e. the fibers of F have dimension 1), an integral manifold for F is nothing but an integral curve from the theory of differential equations, and this curve may wind around X in such a way that its image is dense. A special case of this occurs if we consider the torus as the quotient of the plane by the subgroup generated by the two unit vectors. A straight line with irrational slope in the plane gets mapped on a dense integral curve on the torus.

If Y is a submanifold of X, then of course the inclusion $j: Y \to X$ is an injective immersion, and in this case, the condition that it be an integral manifold for F simply means that $T(Y) = F|Y$ (F restricted to Y).

We now have the local uniqueness of integral manifolds, corresponding to the local uniqueness of integral curves.

Theorem 4.1. *Let* Y, Z *be integral submanifolds of* X *for the subbundle* F *of* TX, *passing through a point* x_0. *Then there exists an open neighborhood* U *of* x_0 *in* X, *such that*

$$Y \cap U = Z \cap U.$$

Proof. Let U be an open neighborhood of x_0 in X such that we have a chart

$$U \rightarrow V \times W$$

with

$$x_0 \mapsto (y_0, w_0),$$

and Y corresponds to all points (y, w_0), $y \in V$. In other words, Y corresponds to a factor in the product in the chart. If V is open in $\mathbf{F}_1$ and W open in $\mathbf{F}_2$, with $\mathbf{F}_1 \times \mathbf{F}_2 = E$, then the subbundle $\mathbf{F}$ is represented by the projection

$$V \times W \times \mathbf{F}_1$$

$$V \times W$$

Shrinking Z, we may assume that $Z \subset U$. Let $h: Z \rightarrow V \times W$ be the restriction of the chart to Z, and let $h = (h_1, h_2)$ be represented by its two components. By assumption, $h'(x)$ maps $\mathbf{E}$ into $\mathbf{F}_1$ for every $x \in Z$. Hence h_2 is constant, so that $h(Z)$ is contained in the factor $V \times \{w_0\}$. It follows at once that $h(Z) = V_1 \times \{w_0\}$ for some open V_1 in V, and we can shrink U to a product $V_1 \times W_1$ (where W_1 is a small open set in W containing w_0) to conclude the proof.

We wish to get a maximal connected integral manifold for an integrable subbundle F of TX passing through a given point, just as we obtained a maximal integral curve. For this, it is just as easy to deal with the nonconnected case, following Chevalley's treatment in his book on *Lie Groups*. (Note the historical curiosity that vector bundles were invented about a year after Chevalley published his book, so that the language of vector bundles, or the tangent bundle, is absent from Chevalley's presentation. In fact, Chevalley used a terminology which now appears terribly confusing for the notion of a tangent subbundle, and it will not be repeated here!)

We give a new manifold structure to X, depending on the integrable tangent subbundle F, and the manifold thus obtained will be denoted by

X_F. This manifold has the same set of points as X. Let $x \in X$. We know from the local uniqueness theorem that a submanifold Y of X which is at the same time an integral manifold for F is locally uniquely determined. A chart for this submanifold locally at x is taken to be a chart for X_F. It is immediately verified that the collection of such charts is an atlas, which defines our manifold X_F. (We lose one order of differentiability.) The identity mapping

$$j: X_F \to X$$

is then obviously an injective immersion, satisfying the following universal properties.

Theorem 4.2. *Let F be an integrable tangent subbundle over X. If*

$$f: Y \to X$$

is a morphism such that $Tf: TY \to TX$ maps TY into F, then the induced map

$$f_F: Y \to X_F$$

(same values as f but viewed as a map into the new manifold X_F) is also a morphism. Furthermore, if f is an injective immersion, then f_F induces an isomorphism of Y onto an open subset of X_F.

Proof. Using the local product structure as in the proof of the local uniqueness Theorem 4.1, we see at once that f_F is a morphism. In other words, locally, f maps a neighborhood of each point of Y into a submanifold of X which is tangent to F. If in addition f is an injective immersion, then from the definition of the charts on X_F, we see that f_F maps Y bijectively onto an open subset of X_F, and is a local isomorphism at each point. Hence f_F induces an isomorphism of Y with an open subset of X_F, as was to be shown.

Corollary 4.3. *Let $X_F(x_0)$ be the connected component of X_F containing a point x_0. If $f: Y \to X$ is an integral manifold for F passing through x_0, and Y is connected, then there exists a unique morphism*

$$h: Y \to X_F(x_0)$$

making the following diagram commutative:

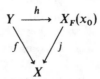

and h induces an isomorphism of Y onto an open subset of $X_F(x_0)$.

Proof. Clear from the preceding discussion.

Note the general functorial behavior of the integral manifold. If

$$g\colon X \to X'$$

is an isomorphism, and F is an integrable tangent subbundle over X, then $F' = (Tg)(F) = g_*F$ is an integrable bundle over X'. Then the following diagram is commutative:

$$
\begin{array}{ccc}
X_F & \xrightarrow{\ g_F\ } & X'_{F'} \\
{\scriptstyle j}\big\downarrow & & \big\downarrow{\scriptstyle j'} \\
X & \xrightarrow{\ \ g\ \ } & X'
\end{array}
$$

The map g_F is, of course, the map having the same values as g, but viewed as a map on the manifold X_F.

VI, §5. LIE GROUPS AND SUBGROUPS

It is not our purpose here to delve extensively into Lie groups, but to lay the groundwork for their theory. For more results, we refer the reader to texts on Lie groups, differential geometry, and also to the paper by W. Graeub [Gr 61].

By a **group manifold**, or a **Lie group** G, we mean a manifold with a group structure, that is a law of composition and inverse,

$$\tau\colon G \times G \to G \qquad \text{and} \qquad G \to G$$

which are morphisms. Thus each $x \in G$ gives rise to a left translation

$$\tau^x\colon G \to G$$

such that $\tau^x(y) = xy$.

When dealing with groups, we shall have to distinguish between isomorphisms in the category of manifolds, and isomorphisms in the category of group manifolds, which are also group homomorphisms. Thus we shall use prefixes, and speak of group manifold isomorphism, or manifold isomorphism as the case may be. We abbreviate these by GM-isomorphism or M-isomorphism. We see that left translation is an M-isomorphism, but not a GM-isomorphism.

Let e denote the origin (unit element) of G. If $v \in T_e G$ is a tangent vector at the origin, then we can translate it, and we obtain a map

$$(x, v) \mapsto \tau^x_* v = \xi_v(x)$$

which is easily verified to be a VB-isomorphism

$$G \times T_e G \to TG$$

from the product bundle to the tangent bundle of G. This is done at once using charts. Recall that $T_e G$ can be viewed as a vector space, using any local trivialization of G at e to get a linear isomorphism of $T_e G$ with the standard space on which G is modeled. Thus we see that the tangent bundle of a Lie group is trivializable.

A vector field ξ over G is called **left invariant** if $\tau_*^x \xi = \xi$ for all $x \in G$. Note that the map

$$x \mapsto \xi_v(x)$$

described above is a left invariant vector field, and that the association

$$v \mapsto \xi_v$$

obviously establishes a linear isomorphism between $T_e G$ and the vector space of left invariant vector fields on G. The space of such vector fields will be denoted by $\mathfrak{g}$ or $\mathfrak{l}(G)$, and will be called the **Lie algebra** of G, because of the following results.

Proposition 5.1. *Let ξ, η be left invariant vector fields on G. Then $[\xi, \eta]$ is also left invariant.*

Proof. This follows from the general functorial formula

$$\tau_*^x [\xi, \eta] = [\tau_*^x \xi, \tau_*^x \eta] = [\xi, \eta].$$

Under the linear isomorphism of $T_e G$ with $\mathfrak{l}(G)$, we can view $\mathfrak{l}(G)$ as a vector space. By a **Lie subalgebra** of $\mathfrak{l}(G)$ we shall mean a closed subspace $\mathfrak{h}$ having the property that if ξ, $\eta \in \mathfrak{h}$, then $[\xi, \eta] \in \mathfrak{h}$ also.

Let G, H be Lie groups. A map

$$f: H \to G$$

will be called a **homomorphism** if it is a group homomorphism and a morphism in the category of manifolds. Such a homomorphism induces a linear map

$$T_e f = f_*: T_e H \to T_e G,$$

and it is clear that it also induces a corresponding linear map

$$\mathfrak{l}(H) \to \mathfrak{l}(G),$$

also denoted by f_*. Namely, if $v \in T_e H$ and ξ_v is the left invariant vector field on H induced by v, then

$$f_* \xi_v = \xi_{f_* v}.$$

The general functorial property of related vector fields applies to this case, and shows that the induced map

$$f_* \colon \mathfrak{l}(H) \to \mathfrak{l}(G)$$

is also a Lie algebra homomorphism, namely for $\xi, \eta \in \mathfrak{l}(H)$ we have

$$f_*[\xi, \eta] = [f_* \xi, f_* \eta].$$

Now suppose that the homomorphism $f \colon H \to G$ is also an immersion at the origin of H. Then by translation, one sees that it is an immersion at every point. If in addition it is an injective immersion, then we shall say that f is a **Lie subgroup** of G. We see that in this case, f induces an injection

$$f_* \colon \mathfrak{l}(H) \to \mathfrak{l}(G).$$

The image of $\mathfrak{l}(H)$ in $\mathfrak{l}(G)$ is a Lie subalgebra of $\mathfrak{l}(G)$.

In general, let $\mathfrak{h}$ be a Lie subalgebra of $\mathfrak{l}(G)$ and let F_e be the corresponding subspace of $T_e G$. For each $x \in G$, let

$$F_x = \tau_*^x F_e.$$

Then F_x is a subspace of $T_x G$, and using local charts, it is clear that the collection $F = \{F_x\}$ is a subbundle of TG, which is left invariant. Furthermore, if

$$f \colon H \to G$$

is a homomorphism which is an injective immersion, and if $\mathfrak{h}$ is the image of $\mathfrak{l}(H)$, then we also see that f is an integral manifold for the subbundle F. We shall now see that the converse holds, using Frobenius' theorem.

Theorem 5.2. *Let G be a Lie group, $\mathfrak{h}$ a Lie subalgebra of $\mathfrak{l}(G)$, and let F be the corresponding left invariant subbundle of TG. Then F is integrable.*

Proof. I owe the proof to Alan Weinstein. It is based on the following lemma.

Lemma 5.3. *Let X be a manifold, let ξ, η be vector fields at a point x_0, and let F be a subbundle of TX. If $\xi(x_0) = 0$ and ξ is contained in F, then $[\xi, \eta](x_0) \in F$.*

Proof. We can deal with the local representations, such that $X = U$ is open in $\mathbf{E}$, and F corresponds to a factor, that is

$$TX = U \times \mathbf{F}_1 \times \mathbf{F}_2 \quad \text{and} \quad F = U \times \mathbf{F}_1.$$

We may also assume without loss of generality that $x_0 = 0$. Then $\xi(0) = 0$, and $\xi\colon U \to \mathbf{F}_1$ may be viewed as a map into $\mathbf{F}_1$. We may write

$$\xi(x) = A(x)x,$$

with a morphism $A\colon U \to L(\mathbf{E}, \mathbf{F}_1)$. Indeed,

$$\xi(x) = \int_0^1 \xi'(tx)\, dt \cdot x,$$

and $A(x) = \mathrm{pr}_1 \circ \int_0^1 \xi'(tx)\, dt$, where pr_1 is the projection on $\mathbf{F}_1$. Then

$$[\xi, \eta](x) = \eta'(x)\xi(x) - \xi'(x)\eta(x)$$
$$= \eta'(x)A(x)x - A'(x) \cdot x \cdot \eta(x) - A(x) \cdot \eta(x),$$

whence

$$[\xi, \eta](0) = A(0)\eta(0).$$

Since $A(0)$ maps $\mathbf{E}$ into $\mathbf{F}_1$, we have proved our lemma.

Back to the proof of the proposition. Let ξ, η be vector fields at a point x_0 in G, both contained in the invariant subbundle F. There exist invariant vector fields ξ_0 and η_0 and x_0 such that

$$\xi(x_0) = \xi_0(x_0) \quad \text{and} \quad \eta(x_0) = \eta_0(x_0).$$

Let

$$\xi_1 = \xi - \xi_0 \quad \text{and} \quad \eta_1 = \eta - \eta_0.$$

Then ξ_1, η_1 vanish at x_0 and lie in F. We get:

$$[\xi, \eta] = \sum_{i,j} [\xi_i, \eta_j].$$

The proposition now follows at once from the lemma.

Theorem 5.4. *Let G be a Lie group, let $\mathfrak{h}$ be a Lie subalgebra of $\mathfrak{l}(G)$, and let F be its associated invariant subbundle. Let*

$$j\colon H \to G$$

be the maximal connected integral manifold of F passing through e. Then H is a subgroup of G, and j: H → G is a Lie subgroup of G. The association between 𝔥 and j: H → G establishes a bijection between Lie subalgebras of l(G) and Lie subgroups of G.

Proof. Let $x \in H$. The M-isomorphism τ^x induces a VB-isomorphism of F onto itself, in other words, F is invariant under τ^x_*. Furthermore, since H passes through e, and xe lies in H, it follows that $j: H \to G$ is also the maximal connected integral manifold of F passing through x. Hence x maps H onto itself. From this we conclude that if $y \in H$, then $xy \in H$, and there exists some $y \in H$ such that $xy = e$, whence $x^{-1} \in H$. Hence H is a subgroup. The other assertions are then clear.

If H is a Lie subgroup of G, belonging to the Lie algebra $\mathfrak{h}$, and F is the associated integrable left invariant tangent subbundle, then the integral manifold for F passing through a given point x is simply the translation xH, as one sees from first functorial principles.

When $\mathfrak{h}$ is 1-dimensional, then it is easy to see that the Lie subgroup is in fact a homomorphic image of an integral curve

$$\alpha: \mathbf{R} \to G$$

which is a homomorphism, and such that $\alpha'(0) = v$ is any vector in $T_e G$ which is the value at e of a non-zero element of $\mathfrak{h}$. Changing this vector merely reparametrizes the curve. The integral curve may coincide with the subgroup, or it comes back on itself, and then the subgroup is essentially a circle. Thus the integral curve need not be equal to the subgroup. However, locally near $t = 0$, they do coincide. Such an integral curve is called a **one-parameter subgroup** of G.

Using Theorem 1.5 of Chapter V, it is then easy to see that if the Lie algebra of a connected Lie group G is commutative, then G itself is commutative. One first proves this for elements in a neighborhood of the origin, using 1-parameter subgroups, and then one gets the statement globally by expressing G as a union of products

$$UU \cdots U,$$

where U is a symmetric connected open neighborhood of the unit element. All of these statements are easy to prove, and belong to the first chapter of a book on Lie groups. Our purpose here is merely to lay the general foundations essentially belonging to general manifold theory.

CHAPTER VII

Metrics

In our discussion of vector bundles, we put no greater structure on the fibers than that of topological vector space (of the same category as those used to build up manifolds). One can strengthen the notion so as to include the metric structure, and we are thus led to consider Hilbert bundles, whose fibers are Hilbert spaces.

Aside from the definitions, and basic properties, we deal with two special topics. On the one hand, we complete our uniqueness theorem on tubular neighborhoods by showing that when a Riemannian metric is given, a tubular neighborhood can be straightened out to a metric one. Secondly, we show how a Riemannian metric gives rise in a natural way to a spray, and thus how one recovers geodesics. The fundamental 2-form is used to identify the vector fields and 1-forms on the tangent bundle, identified with the cotangent bundle by the Riemannian metric.

We assume throughout that our manifolds are sufficiently differentiable so that all our statements make sense. (For instance, when dealing with sprays, we take $p \geq 3$.)

Of necessity, we shall use the standard spectral theorem for (bounded) symmetric operators. A self-contained treatment will be given in the appendix.

VII, §1. DEFINITION AND FUNCTORIALITY

For Riemannian geometry, we shall deal with a euclidean vector space, that is a vector space with a positive definite scalar product.

It turns out that some basic properties have only to do with a weaker property of the space E on which a manifold is modeled, namely that the space **E** is self dual, via a symmetric non-singular bilinear form. Thus we only assume this property until more is needed. We recall that such a form is a bilinear map

$$(v, w) \mapsto \langle v, w \rangle \qquad \text{of} \quad \mathbf{E} \times \mathbf{E} \to \mathbf{R}$$

such that $\langle v, w \rangle = \langle w, v \rangle$ for all $v, w \in \mathbf{E}$, and the corresponding map of **E** into the dual space $L(\mathbf{E})$ is a linear isomorphism.

Examples. Of course, the standard positive definite scalar product on Euclidean space provides the easiest (in some sense) example of a self dual vector space. But the physicists are interested in $\mathbf{R}^4$ with the scalar product such that the square of a vector (x, y, z, t) is $x^2 + y^2 + z^2 - t^2$. This scalar product is non-singular. For one among many nice applications of the indefinite case, cf. for instance [He 84] and [Gu 91], dealing with Huygens' principle.

We consider $L^2_{\text{sym}}(\mathbf{E})$, the vector space of continuous bilinear forms

$$\lambda: \mathbf{E} \times \mathbf{E} \to \mathbf{R}$$

which are symmetric. If x is fixed in **E**, then the linear form $\lambda_x(y) = \lambda(x, y)$ is represented by an element of **E** which we denote by Ax, where A is a linear map of **E** into itself. The symmetry of λ implies that A is symmetric, that is we have

$$\lambda(x, y) = \langle Ax, y \rangle = \langle x, Ay \rangle$$

for all $x, y \in \mathbf{E}$. Conversely, given a symmetric continuous linear map $A: \mathbf{E} \to \mathbf{E}$ we can define a continuous bilinear form on **E** by this formula. Thus $L^2_{\text{sym}}(\mathbf{E})$ is in bijection with the set of such operators, and is itself a vector space, the norm being the usual operator norm. Suppose **E** is a euclidean space, and in particular, **E** is self dual.

The subset of $L^2_{\text{sym}}(\mathbf{E})$ consisting of those forms corresponding to symmetric **positive definite operators** (by definition such that $A \geq \epsilon I$ for some $\epsilon > 0$) will be called the **Riemannian** of **E** and be denoted by Ri(**E**). Forms λ in Ri(**E**) are called positive definite. The associated operator A of such a form is invertible, because its spectrum does not contain 0.

In general, suppose only that **E** is self dual. The space $L^2_{\text{sym}}(\mathbf{E})$ contains as an open subset the set of non-singular symmetric bilinear forms, which we denote by Met(**E**), and which we call the set of **metrics or pseudo Riemannian metrics**. In view of the operations on vector bundles (Chapter III, §4) we can apply the functor L^2_{sym} to any bundle whose fibers are self dual. Thus if $\pi: E \to X$ is such a bundle, then we can form $L^2_{\text{sym}}(\pi)$. A section of $L^2_{\text{sym}}(\pi)$ will be called by definition a **symmetric bilinear form** on π. A **(pseudo Riemannian) metric** on π (or on E) is defined to be a symmetric bilinear form on π, whose image lies in the open set of metrics at each point. We let $\mathrm{Met}(\pi)$ be the set of metrics on π , which we also call the set of metrics on E, and may denote by $\mathrm{Met}(E)$.

If $\mathbf{E}$ is a euclidean space and the image of the section of $L^2_{\mathrm{sym}}(\pi)$ lies in the Riemannian space $\mathrm{Ri}(\pi_x)$ at each point x, in order words, if on the fiber at each point the non-singular symmetric bilinear form is actually positive definite, then we call the metric **Riemannian**. Let us denote a metric by g, so that $g(x) \in \mathrm{Met}(E_x)$ for each $x \in X$, and lies in $\mathrm{Ri}(E_x)$ if the metric is Riemannian. Then $g(x)$ is a non-singular symmetric bilinear form in general, and in the Riemannian case, it is positive definite in addition.

A pair (X, g) consisting of a manifold X and a (pseudo Riemannian) metric g will be called a **pseudo Riemannian manifold**. It will be called a **Riemannian manifold** if the manifold is modeled on a euclidean space, and the metric is Riemannian.

Observe that the sections of $L^2_{\mathrm{sym}}(\pi)$ form an infinite dimensional vector space (abstract) but that the Riemannian metrics do not. They form a convex cone. Indeed, if $a, b > 0$ and g_1, g_2 are two Riemannian metrics, then $ag_1 + bg_2$ is also a Riemannian metric.

Suppose we are given a VB-trivialization of π over an open subset U of X, say

$$\tau\colon \pi^{-1}(U) \to U \times \mathbf{E}.$$

We can transport a given pseudo Riemannian metric g (or rather its restriction to $\pi^{-1}(U)$) to $U \times \mathbf{E}$. In the local representation, this means that for each $x \in U$ we can identify $g(x)$ with a symmetric invertible operator A_x giving rise to the metric. The operator A_x is positive definite in the Riemannian case. Furthermore, the map

$$x \mapsto A_x$$

from U into the vector space $L(\mathbf{E}, \mathbf{E})$ is a morphism.

As a matter of notation, we sometimes write g_x instead of $g(x)$. Thus if v, w are two vectors in E_x, then $g_x(v, w)$ is a number, and is more convenient to write than $g(x)(v, w)$. We shall also write $\langle v, w \rangle_x$ if the metric g is fixed once for all.

Proposition 1.1. *Let X be a manifold admitting partitions of unity. Let $\pi\colon E \to X$ be a vector bundle whose fibers are euclidean vector spaces. Then π admits a Riemannian metric.*

Proof. Find a partition of unity $\{U_i, \varphi_i\}$ such that $\pi|U_i$ is trivial, that is such that we have a trivialization

$$\pi_i\colon \pi^{-1}(U_i) \to U_i \times \mathbf{E}$$

(working over a connected component of X, so that we may assume the fibers toplinearly isomorphic to a fixed space $\mathbf{E}$). We can then find a Riemannian metric on $U_i \times \mathbf{E}$ in a trivial way. By transport of structure, there exists a Riemannian metric g_i on $\pi | U_i$ and we let

$$g = \sum \varphi_i g_i.$$

Then g is a Riemannian metric on x.

Let us investigate the functorial behavior of metrics. Consider a VB-morphism

$$
\begin{array}{ccc}
E' & \xrightarrow{\ f\ } & E \\
{\scriptstyle \pi'}\downarrow & & \downarrow{\scriptstyle \pi} \\
X & \xrightarrow[\ f_0\ ]{} & Y
\end{array}
$$

with vector bundles E' and E over X and Y respectively, whose fibers are self dual spaces. Let g be a symmetric bilinear form on π, so that for each $y \in Y$ we have a bilinear, symmetric map

$$g(y)\colon E_y \times E_y \to \mathbf{R}.$$

Then the composite map

$$E'_x \times E'_x \to E_y \times E_y \to \mathbf{R}$$

with $y = f(x)$ is a symmetric bilinear form on E'_x and one verifies immediately that it gives rise to such a form, on the vector bundle π', which will be denoted by $f^*(g)$. Then f induces a map

$$L^2_{\mathrm{sym}}(f) = f^*\colon\ L^2_{\mathrm{sym}}(\pi) \to L^2_{\mathrm{sym}}(\pi').$$

Furthermore, if f_x is injective and splits for each $x \in X$, and g is a metric (resp. g is a Riemannian metric in the euclidean case), then obviously so is $f^*(g)$, and we can view f^* as mapping $\mathrm{Met}(\pi)$ into $\mathrm{Met}(\pi')$ (resp. $\mathrm{Ri}(\pi)$ into $\mathrm{Ri}(\pi')$ in the Riemannian case).

Let X be a manifold modeled on a euclidean space and let $T(X)$ be its tangent bundle. By abuse of language, we call a metric on $T(X)$ also a metric on X and write $\mathrm{Met}(X)$ instead of $\mathrm{Met}(T(X))$. Similarly, we write $\mathrm{Ri}(X)$ instead of $\mathrm{Ri}(T(X))$.

Let $f\colon X \to Y$ be an immersion. Then for each $x \in X$, the linear map

$$T_x f\colon\ T_x(X) \to T_{f(x)}(Y)$$

is injective, and splits, and thus we obtain a contravariant map

$$f^*\colon \operatorname{Ri}(Y) \to \operatorname{Ri}(X),$$

each Riemannian metric on Y inducing a Riemannian metric on X.

A similar result applies in the pseudo Riemannian case. If (Y, g) is Riemannian, and f is merely of class C^1 but not necessarily an immersion, then the pull back $f^*(g)$ is not necessarily positive definite, but is merely what we call **semipositive**. In general, if (X, h) is pseudo Riemannian and $h(v, v) \geqq 0$ for all $v \in T_x X$, all x, then (X, h) is called **semi Riemannian**. Thus the pull back of a semi Riemannian metric is semi Riemannian.

The next five sections will be devoted to considerations which apply specifically to the Riemannian case, where positivity plays a central role.

VII, §2. THE METRIC GROUP

Let $\mathbf{E}$ be a euclidean vector space. The group of linear automorphisms $\operatorname{Laut}(\mathbf{E})$ contains the group $\operatorname{Maut}(\mathbf{E})$ of metric automorphisms, that is those linear automorphisms which preserve the inner product:

$$\langle Av, Aw \rangle = \langle v, w \rangle$$

for all $v, w \in \mathbf{E}$. We note that A is metric if and only if $A^* A = I$.

As usual, we say that a linear map $A\colon \mathbf{E} \to \mathbf{E}$ is **symmetric** if $A^* = A$ and that it is **skew-symmetric** if $A^* = -A$. We have a direct sum decomposition of the space $L(\mathbf{E}, \mathbf{E})$ in terms of the two closed subspaces of symmetric and skew-symmetric operators:

$$A = \tfrac{1}{2}(A + A^*) + \tfrac{1}{2}(A - A^*).$$

We denote by $\operatorname{Sym}(\mathbf{E})$ and $\operatorname{Sk}(\mathbf{E})$ the vector spaces of symmetric and skew-symmetric maps respectively. The word **operator** will always mean linear map of $\mathbf{E}$ into itself.

Proposition 2.1. *For all operators A, the series*

$$\exp(A) = I + A + \frac{A^2}{2!} + \cdots$$

converges. If A commutes with B, then

$$\exp(A + B) = \exp(A)\exp(B).$$

For all operators sufficiently close to the identity I, the series

$$\log(A) = \frac{(A - I)}{1} + \frac{(A - I)^2}{2} + \cdots$$

converges, and if A commutes with B, then

$$\log(AB) = \log(A) + \log(B).$$

Proof. Standard.

We leave it as an exercise to the reader to show that the exponential function gives a C^∞-morphism of $L(\mathbf{E}, \mathbf{E})$ into itself. Similarly, a function admitting a development in power series say around 0 can be applied to the set of operators whose bound is smaller than the radius of convergence of the series, and gives a C^∞-morphism.

Proposition 2.2. *If A is symmetric (resp. skew-symmetric), then* exp(A) *is symmetric positive definite (resp. metric). If A is a linear automorphism sufficiently close to I and is positive definite symmetric (resp. metric), then* log(A) *is symmetric (resp. skew-symmetric).*

Proof. The proofs are straightforward. As an example, let us carry out the proof of the last statement. Suppose A is Hilbertian and sufficiently close to I. Then $A^*A = I$ and $A^* = A^{-1}$. Then

$$\log(A)^* = \frac{(A^* - I)}{1} + \cdots$$

$$= \log(A^{-1}).$$

If A is close to I, so is A^{-1}, so that these statements make sense. We now conclude by noting that $\log(A^{-1}) = -\log(A)$. All the other proofs are carried out in a similar fashion, taking a star operator in series term by term, under conditions which insure convergence.

The exponential and logarithm functions give inverse C^∞ mappings between neighborhoods of 0 in $L(\mathbf{E}, \mathbf{E})$ and neighborhoods of I in Laut($\mathbf{E}$). Furthermore, the direct sum decomposition of $L(\mathbf{E}, \mathbf{E})$ into symmetric and skew-symmetric subspaces is reflected locally in a neighborhood of I by a C^∞ direct product decomposition into positive definite and metric automorphisms. This direct product decomposition can be translated multiplicatively to any linear automorphism, because if $A \in \text{Laut}(\mathbf{E})$ and B is close to A, then

$$B = AA^{-1}B = A(I - (I - A^{-1}B))$$

and $(I - A^{-1}B)$ is small. This proves:

Proposition 2.3. *The group of metric automorphisms* Maut(**E**) *of* **E** *is a closed submanifold of* Laut(**E**).

In addition to this local result, we get a global one also:

Proposition 2.4. *The exponential map gives a* C^∞-*isomorphism from the space* Sym(**E**) *of symmetric endomorphisms of* **E** *and the space* Pos(**E**) *of symmetric positive definite automorphisms of* **E**.

Proof. We must construct its inverse, and for this we use the spectral theorem. Given A, symmetric positive definite, the analytic function log t is defined on the spectrum of A, and thus log A is symmetric. One verifies immediately that it is the inverse of the exponential function (which can be viewed in the same way). We can expand log t around a large positive number c, in a power series uniformly and absolutely convergent in an interval $0 < \epsilon \leqq t \leqq 2c - \epsilon$, to achieve our purposes.

Proposition 2.5. *The manifold of linear automorphisms of the Euclidean space* **E** *is* C^∞-*isomorphic to the product of the metric automorphisms and the positive definite symmetric automorphisms, under the mapping*

$$\text{Maut}(\mathbf{E}) \times \text{Pos}(\mathbf{E}) \to \text{Laut}(\mathbf{E})$$

given by

$$(H, P) \to HP.$$

Proof. Our map is induced by a continuous bilinear map of

$$L(\mathbf{E}, \mathbf{E}) \times L(\mathbf{E}, \mathbf{E})$$

into $L(\mathbf{E}, \mathbf{E})$ and so is C^∞. We must construct an inverse, or in other words express any given linear automorphism A in a unique way as a product $A = HP$ where H is metric, P is symmetric positive definite, and both H, P depend C^∞ on A. This is done as follows. First we note that A^*A is symmetric positive definite (because $\langle A^*Av, v \rangle = \langle Av, Av \rangle$, and furthermore, A^*A is a linear automorphism. By linear algebra, A^*A can be diagonalized. We let

$$P = (A^*A)^{1/2}$$

and let $H = AP^{-1}$. Then H is metric, because

$$H^*H = (P^{-1})^*A^*AP^{-1} = I.$$

Both P and H depend differentiably on A since all constructions involved are differentiable.

There remains to be shown that the expression as a product is unique. If $A = H_1 P_1$ where H_1, P_1 are metric and symmetric positive definite respectively, then

$$H^{-1}H_1 = PP_1^{-1},$$

and we get $H_2 = PP_1^{-1}$ for some metric automorphism H_2. By definition,

$$I = H_2^* H_2 = (PP_1^{-1})^* PP_1^{-1}$$

and from the fact that $P^* = P$ and $P_1^* = P_1$, we find

$$P^2 = P_1^2.$$

Taking the log, we find $2 \log P = 2 \log P_1$. We now divide by 2 and take the exponential, thus giving $P = P_1$ and finally $H = H_1$. This proves our proposition.

VII, §3. REDUCTION TO THE METRIC GROUP

We define a new category of bundles, namely the **metric bundles** over X, denoted by $\mathrm{MB}(X)$. As before, we would denote by $\mathrm{MB}(X, \mathbf{E})$ those metric bundles whose fiber is a euclidean space $\mathbf{E}$.

Let $\pi: E \to X$ be a vector bundle over X, and assume that it has a trivialization $\{(U_i, \tau_i)\}$ with trivializing maps

$$\tau_i: \pi^{-1}(U_i) \to U_i \times \mathbf{E}$$

where $\mathbf{E}$ is a euclidean space, such that each linear automorphism $(\tau_j \tau_i^{-1})_x$ is a metric automorphism. Equivalently, we could also say that τ_{ix} is a metric isomorphism. Such a trivialization will be called a **metric trivialization**. Two such trivializations are called **metric-compatible** if their union is again a metric trivialization. An equivalence class of such compatible trivializations constitutes what we call a **metric bundle** over X. Any such metric bundle determines a unique vector bundle, simply by taking the VB-equivalence class determined by the trivialization.

Given a metric trivialization $\{(U_i, \tau_i)\}$ of a vector bundle π over X, we can define on each fiber π_x a euclidean structure. Indeed, for each x we select an open set U_i in which x lies, and then transport to π_x the scalar product in $\mathbf{E}$ by means of τ_{ix}. By assumption, this is independent of the choice of U_i in which x lies. Thus in a metric bundle, we can assume that the fibers are metric spaces.

It is perfectly possible that several distinct metric bundles determine the same vector bundle.

Any metric bundle determining a given vector bundle π will be said to be a **reduction of π to the metric group**.

We can make metric bundles into a category, if we take for the **MB-morphisms** the VB-morphisms which are injective at each point, and which preserve the metric, again at each point.

Each reduction of a vector bundle to the metric group determines a Riemannian metric on the bundle. Indeed, defining for each $z \in X$ and $v, w \in \pi_x$ the scalar product

$$g_x(v, w) = \langle \tau_{ix}v, \tau_{ix}w \rangle$$

with any metric-trivializing map τ_{ix} such that $x \in U_i$, we get a morphism

$$x \mapsto g_x$$

of X into the sections of $L^2_{\text{sym}}(\pi)$ which are positive definite. We also have the converse.

Theorem 3.1. *Let π be a vector bundle over a manifold X, and assume that the fibers of π are all linearly isomorphic to a euclidean space* **E.** *Then the above map, from reductions of π to the metric group, into the Riemannian metrics, is a bijection.*

Proof. Suppose that we are given an ordinary VB-trivialization $\{(U_i, \tau_i)\}$ of π. We must construct an MB-trivialization. For each i, let g_i be the Riemannian metric on $U_i \times \mathbf{E}$ transported from $\pi^{-1}(U_i)$ by means of τ_i. Then for each $x \in U_i$, we have a positive definite symmetric operator A_{ix} such that

$$g_{ix}(v, w) = \langle A_{ix}v, w \rangle$$

for all $v, w \in \mathbf{E}$. Let B_{ix} be the square root of A_{ix}. We define the trivialization σ_i by the formula

$$\sigma_{ix} = B_{ix}\tau_{ix}$$

and contend that $\{(U_i, \sigma_i)\}$ is a metric trivialization. Indeed, from the definition of g_{ix}, it suffices to verify that the VB-isomorphism

$$B_i \colon U_i \times \mathbf{E} \to U_i \times \mathbf{E}$$

given by B_{ix} on each fiber, carries g_i on the usual metric. But we have, for

$v, w \in E$:

$$\langle B_{ix}v, \ B_{ix}w \rangle = \langle A_{ix}v, \ w \rangle$$

since B_{ix} is symmetric, and equal to the square root of A_{ix}. This proves what we want.

At this point, it is convenient to make an additional comment on normal bundles.

Let α, β be two metric bundles over the manifold X, and let $f: \alpha \to \beta$ be an MB-morphism. Assume that

$$0 \to \alpha \xrightarrow{f} \beta$$

is exact. Then by using the Riemannian metric, there is a natural way of constructing a splitting for this sequence (cf. Chapter III, §5).

By elementary linear algebra, if $\mathbf{F}$ is a subspace of a euclidean space, then $\mathbf{E}$ is the direct sum

$$\mathbf{E} = \mathbf{F} \oplus \mathbf{F}^{\perp}$$

of $\mathbf{F}$ and its orthogonal complement, consisting of all vectors perpendicular to $\mathbf{F}$.

In our exact sequence, we may view f as an injection. For each x we let $\alpha_x^{\perp}$ be the orthogonal complement of α_x in β_x. Then we shall find an exact sequence of VB-morphisms

$$\beta \xrightarrow{h} \alpha \to 0$$

whose kernel is $\alpha^{\perp}$ (set theoretically). In this manner, the collection of orthogonal complements $\alpha_x^{\perp}$ can be given the structure of a metric bundle.

For each x we can write $\beta_x = \alpha_x \oplus \alpha_x^{\perp}$ and we define h_x to be the projection in this direct sum decomposition. This gives us a mapping $h: \beta \to \alpha$, and it will suffice to prove that h is a VB-morphism. In order to do this, we may work locally. In that case, after taking suitable VB-automorphisms over a small open set U of X, we can assume that we deal with the following situation.

Our vector bundle β is equal to $U \times \mathbf{E}$ and α is equal to $U \times \mathbf{F}$ for some subspace $\mathbf{F}$ of $\mathbf{E}$, so that we can write $\mathbf{E} = \mathbf{F} \times \mathbf{F}^{\perp}$. Our MB-morphism is then represented for each x by an injection $f_x: \mathbf{F} \to \mathbf{E}$:

$$U \times \mathbf{F} \xrightarrow{f} U \times \mathbf{E}.$$

By the definition of exact sequences, we can find two VB-isomorphisms τ and σ such that the following diagram is commutative:

$$
\begin{array}{ccc}
U \times \mathbf{F} & \xrightarrow{\,f\,} & U \times \mathbf{E} \\
\sigma \downarrow & & \downarrow \tau \\
U \times \mathbf{F} & \longrightarrow & U \times \mathbf{E}
\end{array}
$$

and such that the bottom map is simply given by the ordinary inclusion of $\mathbf{F}$ in $\mathbf{E}$. We can transport the Riemannian structure of the bundles on top to the bundles on the bottom by means of σ^{-1} and τ^{-1} respectively. We are therefore reduced to the situation where f is given by the simple inclusion, and the Riemannian metric on $U \times \mathbf{E}$ is given by a family A_x of symmetric positive definite operators on $\mathbf{E}$ $(x \in U)$. At each point x, we have $\langle v, w \rangle_x = \langle A_x v, w \rangle$. We observe that the map

$$
A \colon U \times \mathbf{E} \to U \times \mathbf{E}
$$

given by A_x on each fiber is a VB-automorphism of $U \times \mathbf{E}$. Let $\mathrm{pr}_{\mathbf{F}}$ be the projection of $U \times \mathbf{E}$ on $U \times \mathbf{F}$. It is a VB-morphism. Then the composite

$$
h = \mathrm{pr}_{\mathbf{F}} \circ A
$$

gives us a VB-morphism of $U \times \mathbf{E}$ on $U \times \mathbf{F}$, and the sequence

$$
U \times \mathbf{E} \xrightarrow{\,h\,} U \times \mathbf{F} \to 0
$$

is exact. Finally, we note that the kernel of h consists precisely of the orthogonal complement of $U \times \mathbf{F}$ in each fiber. This proves what we wanted.

VII, §4. METRIC TUBULAR NEIGHBORHOODS

Let $\mathbf{E}$ be a euclidean space. Then the open ball of radius 1 is isomorphic to $\mathbf{E}$ itself under the mapping

$$
v \mapsto \frac{v}{(1 - |v|^2)^{1/2}},
$$

the inverse mapping being

$$
w \mapsto \frac{w}{(1 + |w|^2)^{1/2}}.
$$

If $a > 0$, then any ball of radius a is isomorphic to the unit ball under multiplication by the scalar a (or a^{-1}).

Let X be a manifold, and $\sigma: X \to \mathbf{R}$ a function (morphism) such that $\sigma(x) > 0$ for all $x \in X$. Let $\pi: E \to X$ be a metric bundle over X. We denote by $E(\sigma)$ the subset of E consisting of those vectors v such that, if v lies in E_x, then

$$|v|_x < \sigma(x).$$

Then $E(\sigma)$ is an open neighborhood of the zero section.

Proposition 4.1. *Let X be a manifold and $\pi: E \to X$ a metric bundle. Let $\sigma: X \to \mathbf{R}$ be a morphism such that $\sigma(x) > 0$ for all x. Then the mapping*

$$w \to \frac{\sigma(\pi w)w}{(1 + |w|^2)^{1/2}}$$

gives an isomorphism of E onto $E(\sigma)$.

Proof. Obvious. The inverse mapping is constructed in the obvious way.

Corollary 4.2. *Let X be a manifold admitting partitions of unity, and let $\pi: E \to X$ be a metric bundle over X. Then E is compressible.*

Proof. Let Z be an open neighborhood of the zero section. For each $x \in X$, there exists an open neighborhood V_x and a number $a_x > 0$ such that the vectors in $\pi^{-1}(V_x)$ which are of length $< a_x$ lie in Z. We can find a partition of unity $\{(U_i, \varphi_i)\}$ on X such that each U_i is contained in some $V_{x(i)}$. We let σ be the function

$$\sum a_{x(i)}\varphi_i.$$

Then $E(\sigma)$ is contained in Z, and our assertion follows from the proposition.

Proposition 4.3. *Let X be a manifold. Let $\pi: E \to X$ and $\pi_1: E_1 \to X$ be two metric bundles over X. Let*

$$\lambda: E \to E_1$$

be a VB-isomorphism. Then there exists an isotopy of VB-isomorphisms

$$\lambda_t: E \to E_1$$

with proper domain $[0, 1]$ such that $\lambda_1 = \lambda$ and λ_0 is an MB-isomorphism.

Proof. We find reductions of E and E_1 to the metric group, with metric trivializations $\{(U_i, \tau_i)\}$ for E and $\{(U_i, \rho_i)\}$ for E_1. We can then factor $\rho_i \lambda \tau_i^{-1}$ as in Proposition 2.5, applied to each fiber map:

$$
\begin{array}{ccccc}
U_i \times \mathbf{E} & \longrightarrow & U_i \times \mathbf{E} & \longrightarrow & U_i \times \mathbf{E} \\
\uparrow{\scriptstyle \tau_i} & & \uparrow{\scriptstyle \tau_i} & & \uparrow{\scriptstyle \rho_i} \\
\pi^{-1}(U_i) & \xrightarrow{\ \lambda_P\ } & \pi(U_i^{-1}) & \xrightarrow{\ \lambda_H\ } & \pi_1^{-1}(U_i)
\end{array}
$$

and obtain a factorization of λ into $\lambda = \lambda_H \lambda_P$ where λ_H is a MB-isomorphism and λ_P is a positive definite symmetric VB-automorphism. The latter form a convex set, and our isotopy is simply

$$\lambda_t = \lambda_H \circ \big(tI + (1 + t)\lambda_P \big).$$

(Smooth out the end points if you wish.)

Theorem 4.4. *Let X be a submanifold of Y. Let $\pi: E \to X$ and $\pi_1: E_1 \to X$ be two metric bundles. Assume that E is compressible. Let $f: E \to Y$ and $g: E_1 \to Y$ be two tubular neighborhoods of X in Y. Then there exists an isotopy*

$$f_t: E \to Y$$

of tubular neighborhoods with proper domain $[0, 1]$ and there exists an MB-isomorphism $\mu: E \to E_1$ such that $f_1 = f$ and $f_0 = g\mu$.

Proof. From Theorem 6.2 of Chapter IV, we know already that there exists a VB-isomorphism λ such that $f \approx g\lambda$. Using the preceding proposition, we know that $\lambda \approx \mu$ where μ is a MB-isomorphism. Thus $g\lambda \approx g\mu$ and by transitivity, $f \approx \mu$, as was to be shown.

Remark. In view of Proposition 4.1, we could of course replace the condition that E be compressible by the more useful condition (in practice) that X admit partitions of unity.

VII, §5. THE MORSE LEMMA

Let U be an open set in some euclidean space $\mathbf{E}$, and let f be a C^{p+2} function on U, with $p \geq 1$. We say that x_0 is a **critical point** for f if $Df(x_0) = 0$. We wish to investigate the behavior of f at a critical point. After translations, we can assume that $x_0 = 0$ and that $f(x_0) = 0$. We observe that the second derivative $D^2 f(0)$ is a continuous bilinear form on

E. Let $\lambda = D^2 f(0)$, and for each $x \in \mathbf{E}$ let λ_x be the functional such that $y \mapsto \lambda(x, y)$. If the map $x \mapsto \lambda_x$ is a linear isomorphism of $\mathbf{E}$ with its dual space $\mathbf{E}^\vee$, then we say that λ is **non-singular**, and we say that the critical point is **non-degenerate**.

We recall that a local C^p-isomorphism φ at 0 is a C^p-invertible map defined on an open set containing 0.

Theorem 5.1. *Let f be a C^{p+2} function defined on an open neighborhood of 0 in the euclidean space $\mathbf{E}$, with $p \geq 1$. Assume that $f(0) = 0$, and that 0 is a non-degenerate critical point of f. Then there exists a local C^p-isomorphism at 0, say φ, and an invertible symmetric operator A such that*

$$f(x) = \langle A\varphi(x), \varphi(x) \rangle.$$

Proof. We may assume that U is a ball around 0. We have

$$f(x) = f(x) - f(0) = \int_0^1 Df(tx)x \, dt,$$

and applying the same formula to Df instead of f, we get

$$f(x) = \int_0^1 \int_0^1 D^2 f(stx)tx \cdot x \, ds \, dt = g(x)(x, x)$$

where

$$g(x) = \int_0^1 \int_0^1 D^2 f(stx)t \, ds \, dt.$$

Then g is a C^p map into the vector space of bilinear maps on $\mathbf{E}$, and even the space of symmetric such maps. We know that this vector space is linearly isomorphic to the space of symmetric operators on $\mathbf{E}$, and thus we can write

$$f(x) = \langle A(x)x, x \rangle$$

where $A: U \to \mathrm{Sym}(\mathbf{E})$ is a C^p map of U into the space of symmetric operators on E. A straightforward computation shows that

$$D^2 f(0)(v, w) = \langle A(0)v, w \rangle.$$

Since we assumed that $D^2 f(0)$ is non-singular, this means that $A(0)$ is invertible, and hence $A(x)$ is invertible for all x sufficiently near 0.

Theorem 5.1 is then a consequence of the following result, which expresses locally the uniqueness of a non-singular symmetric form.

Theorem 5.2. *Let* $A\colon U \to \mathrm{Sym}(\mathbf{E})$ *be a* C^p *map of* U *into the open set of invertible symmetric operators on* $\mathbf{E}$. *Then there exists a* C^p *isomorphism of an open subset* U_1 *containing* 0, *of the form*

$$\varphi(x) = C(x)x, \qquad \text{with a } C^p \text{ map} \quad C\colon U_1 \to \mathrm{Laut}(\mathbf{E})$$

such that

$$\langle A(x)x,\, x \rangle = \langle A(0)\varphi(x),\, \varphi(x) \rangle = \langle A(0)C(x)x,\, C(x)x \rangle.$$

Proof. We seek a map C such that

$$C(x)^* A(0) C(x) = A(x).$$

If we let $B(x) = A(0)^{-1} A(x)$, then $B(x)$ is close to the identity I for small x. The square root function has a power series expansion near 1, which is a uniform limit of polynomials, and is C^∞ on a neighborhood of I, and we can therefore take the square root of $B(x)$, so that we let

$$C(x) = B(x)^{1/2}.$$

We contend that this $C(x)$ does what we want. Indeed, since both $A(0)$ and $A(x)$ $\big(\text{or } A(x)^{-1}\big)$ are self-adjoint, we find that

$$B(x)^* = A(x)A(0)^{-1},$$

whence

$$B(x)^* A(0) = A(0)B(x).$$

But $C(x)$ is a power series in $I - B(x)$, and $C(x)^*$ is the same power series in $I - B(x)^*$. The preceding relation holds if we replace $B(x)$ by any power of $B(x)$ (by induction), hence it holds if we replace $B(x)$ by any polynomial in $I - B(x)$, and hence finally, it holds if we replace $B(x)$ by $C(x)$, and thus

$$C(x)^* A(0) C(x) = A(0)C(x)C(x) = A(0)B(x) = A(x).$$

which is the desired relation.

All that remains to be shown is that φ is a local C^p-isomorphism at 0. But one verifies that in fact, $D\varphi(0) = C(0)$, so that what we need follows from the inverse mapping theorem. This concludes the proof of Theorems 5.1 and 5.2.

Corollary 5.3. *Let* f *be a* C^{p+2} *function near* 0 *on the euclidean space* $\mathbf{E}$, *such that* 0 *is a non-degenerate critical point. Then there exists a local* C^p-*isomorphism* ψ *at* 0, *and an orthogonal decomposition* $\mathbf{E} = \mathbf{F} + \mathbf{F}^{\perp}$, *such that if we write* $\psi(x) = y + z$ *with* $y \in \mathbf{F}$ *and* $z \in \mathbf{F}^\perp$, *then*

$$f(\psi(x)) = \langle y, y \rangle - \langle z, z \rangle.$$

Proof. On a space where A is positive definite, we can always make the toplinear isomorphism $x \mapsto A^{1/2}x$ to get the quadratic form to become the given hermitian product $\langle , \rangle$, and similarly on a space where A is negative definite. In general, we decompose $\mathbf{E}$ into a direct orthogonal sum such that the restriction of A to the factors is positive definite and negative definite respectively.

Note. The Morse lemma was proved originally by Morse in the finite dimensional case, using the Gram–Schmidt orthogonalization process. The above proof is due to Palais [Pa 69]. It shows (in the language of coordinate systems) that a function near a critical point can be expressed as a quadratic form after a suitable change of coordinate system (satisfying requirements of differentiability).

VII, §6. THE RIEMANNIAN DISTANCE

Let (X, g) *be a Riemannian manifold.* For each C^1 curve

$$\gamma\colon [a, b] \to X$$

we define its **length**

$$L_g(\gamma) = L(\gamma) = \int_a^b \langle \gamma'(t), \gamma'(t) \rangle_g^{1/2} \, dt = \int_a^b \|\gamma'(t)\|_g \, dt.$$

The norm is the one associated with the positive definite scalar product, i.e. the euclidean norm at each point. We can extend the length to piecewise C^1 paths by taking the sum over the C^1 curves constituting the path. *We assume that X is connected, which is equivalent to the property that any two points can be joined by a piecewise C^1 path.* (If X is connected, then the set of points which can be joined to a given point x_0 by a piecewise C^1 path is immediately verified to be open and closed, so equal to X. The converse, that pathwise connectedness implies connectedness, is even more obvious.)

We define the *g*-**distance** on X for any two points $x, y \in X$ by:

$\mathrm{dist}_g(x, y) = $ greatest lower bound of $L(\gamma)$ for paths γ in X joining x and y.

When g is fixed throughout, we may omit g from the notation and write simply $\mathrm{dist}(x, y)$. It is clear that dist_g is a semidistance, namely it is symmetric in (x, y) and satisfies the triangle inequality. To prove that it is a distance, we have to show that if $x \neq y$ then $\mathrm{dist}_g(x, y) > 0$. In a chart, there is a neighborhood U of x which contains a closed ball $\bar{\mathbf{B}}(x, r)$ with $r > 0$, and such that y lies outside this closed ball. Then any path between x and y has to cross the sphere $\mathbf{S}(x, r)$. *Here we are using the euclidean norm in the chart.* We can also take r so small that the norm in the chart is given by

$$\langle v, w \rangle_{g(x)} = \langle v, A(x)w \rangle,$$

for $v, w \in \mathbf{E}$, and $x \mapsto A(x)$ is a morphism from U into the set of invertible symmetric positive definite operators, such that there exist a number $C_1 > 0$ for which

$$A(x) \geq C_1 I \qquad \text{for all } x \in \bar{\mathbf{B}}(x, r).$$

We then claim that there exists a constant $C > 0$ depending only on r, such that for any piecewise C^1 path γ between x and a point on the sphere $\mathbf{S}(x, r)$ we have

$$L(\gamma) \geq Cr.$$

This will prove that $\mathrm{dist}_g(x, y) \geq Cr > 0$, and will conclude the proof that dist_g is a distance.

By breaking up the path into a sum of C^1 curves, we may assume without loss of generality that our path is such a curve. Furthermore, we may take the interval $[a, b]$ on which γ is defined to be such that $\gamma(b)$ is the first point such that $\gamma(t)$ lies on $\mathbf{S}(x, r)$, and otherwise $\gamma(t) \in \bar{\mathbf{B}}(x, r)$ for $t \in [a, b]$. Let $\gamma(b) = ru$, where u is a unit vector. Write $\mathbf{E}$ as an orthogonal direct sum

$$\mathbf{E} = \mathbf{R}u \perp \mathbf{F},$$

where $\mathbf{F}$ is a subspace. Then $\gamma(t) = s(t)u = w(t)$ with $|s(t)| \leq r$, $s(a) = 0$, $s(b) = r$ and $w(t) \in \mathbf{F}$. Then

$$L(\gamma) = \int_a^b \|\gamma'(t)\|_g \, dt = \int_a^b \langle \gamma'(t), A(\gamma(t))\gamma'(t) \rangle^{1/2} \, dt$$

$$\geq C_1^{1/2} \int_a^b \langle \gamma'(t), \gamma'(t) \rangle^{1/2} \, dt$$

$$\geq C_1^{1/2} \int_a^b |s'(t)| \, dt \quad \text{by Pythagoras}$$

$$\geq C_1^{1/2} r$$

as was to be shown.

In addition, the above local argument also proves:

Proposition 6.1. *The distance* dist_g *defines the given topology on* X. *Equivalently, a sequence* $\{x_n\}$ *in* X *converges to a point* x *in the given topology if and only if* $\text{dist}_g(x_n, x)$ *converges to* 0.

We conclude this section with some remarks on reparametrization. Let

$$\gamma: [a, b] \to X$$

be a piecewise C^1 path in X. To reparametrize γ, we may do so on each subinterval where γ is actually C^1, so assume γ is C^1. Let

$$\varphi: [c, d] \to [a, b]$$

be a C^1 map such that $\varphi(c) = a$ and $\varphi(d) = b$. Then $\gamma \circ \varphi$ is C^1, and is called a **reparametrization** of γ. The chain rule shows that

$$L(\gamma \circ \varphi) = L(\gamma).$$

Define the function $s: [a, b] \to \mathbf{R}$ by

$$s(t) = \int_a^t \|\gamma(t)\|_g \, dt, \qquad \text{so} \quad s(b) = L = L(\gamma).$$

Then s is monotone and $s(a) = 0$, while $s(b) = L(\gamma)$. Suppose that there is only a finite number of values $t \in [a, b]$ such that $\gamma'(t) = 0$. We may then break up $[a, b]$ into subintervals where $\gamma'(t) \neq 0$ except at the end points of the subintervals. Consider each subinterval separately, and say

$$a < a_1 < b_1 < b$$

with $\gamma'(t) \neq 0$ for $t \in (a_1, b_1)$. Let $s(a_1)$ be the length of the curve over the interval $[a, a_1]$. Define

$$s(t) = s(a_1) + \int_{a_1}^t \|\gamma'(t)\|_g \, dt \qquad \text{for} \quad a_1 \leq t \leq b_1.$$

Then s is strictly increasing, and therefore the inverse function $t = \varphi(s)$ is defined over the interval. Thus we can reparametrize the curve by the variable s over the interval $a_1 \leq t \leq b_1$, with the variable s satisfying

$$s(a_1) \leq s \leq s(b_1).$$

Thus the whole path γ on $[a, b]$ is reparametrized by another path

$$\gamma \circ \varphi \colon [0, L] \to X$$

via a piecewise map $f \colon [0, L] \to [a, b]$, such that

$$\|(\gamma \circ \varphi)'(s)\|_g = 1 \quad \text{and} \quad L_0^s(\gamma \circ \varphi) = s.$$

We now define a path $\gamma \colon [a, b] \to X$ to be **parametrized by arc length** if $\|\gamma'(t)\|_g = 1$ for all $t \in [a, b]$. We see that starting with any path γ, with the condition that there is only a finite number of points where $\gamma'(t) = 0$ for convenience, there is a reparametrization of the path by arc length.

Let $f \colon Y \to X$ be a C^p map with $p \geqq 1$. We shall deal with several notions of isomorphisms in different categories, so in the C^p category, we may call f a differential morphism. Suppose (X, g) and (Y, h) are Riemannian manifolds. We say that f is an **isometry**, or a **differential metric isomorphism** if f is a differential isomorphism and $f^*(g) = h$. If f is an isometry, then it is immediate that f preserves distances, i.e. that

$$\text{dist}_g\big(f(y_1), f(y_2)\big) = \text{dist}_h(y_1, y_2) \quad \text{for all} \quad y_1, y_2 \in Y.$$

Note that there is another circumstance of interest with somewhat weaker conditions when $f \colon Y \to X$ is an immersion, so induces an injection $Tf(y) \colon T_y Y \to T_{f(y)} X$ for every $y \in Y$, and we can speak of f being a metric immersion if $f^*(g) = h$. It may even happen that f is a local differential isomorphism at each point of y, as for instance if f is covering map. In such a case, f may be a local isometry, but not a global one, whereby f may not preserve distances on all of Y, possibly because two points $y_1 \neq y_2$ may have the same image $f(y_1) = f(y_2)$.

VII, §7. THE CANONICAL SPRAY

We now come back to the pseudo Riemannian case.

Let X be a pseudo Riemannian manifold, modeled on the self dual space $\mathbf{E}$. The scalar product $\langle , \rangle$ in $\mathbf{E}$ identifies $\mathbf{E}$ with its dual $\mathbf{E}^\vee$. The metric on X gives a linear isomorphism of each tangent space $T_x(X)$ with $T_x^\vee(X)$. If we work locally with $X = U$ open in $\mathbf{E}$ and we make the identification

$$T(U) = U \times \mathbf{E} \quad \text{and} \quad T^\vee(U) = U \times \mathbf{E}^\vee \approx T(U)$$

then the metric gives a VB-isomorphism

$$h \colon T(U) \to T(U)$$

by means of a morphism

$$g: \ U \to L(\mathbf{E}, \mathbf{E})$$

such that $h(x, v) = \big(x, g(x)v\big)$. (With respect to an orthonormal basis, $g(x)$ is represented by a symmetric matrix $\big(g_{ij}(x)\big)$, so the notation here fits what's in other books with their g_{ij}.) The scalar product of the metric at each point x is then given by the formula

$$\langle v, w \rangle_x = \langle v, g(x)w \rangle = \langle g(x)v, w \rangle \qquad \text{for} \quad v, w \in \mathbf{E}.$$

For each $x \in U$ we note that $g'(x)$ maps $\mathbf{E}$ into $L(\mathbf{E}, \mathbf{E})$. For $x \in U$ and $u, v \in E$ we write

$$\big(g'(x)u\big)(v) = g'(x)u \cdot v = g'(x)(u, v).$$

From the symmetry of g, differentiating the symmetry relation of the scalar product, we find that for all $u, v, w \in \mathbf{E}$,

$$\langle g'(x)u \cdot w, v \rangle = \langle g'(x)u \cdot v, w \rangle.$$

So we can interchange the last two arguments in the scalar product without changing the value.

Observe that locally, the tangent linear map

$$T(h): \ T\big(T(U)\big) \to T\big(T(U)\big)$$

is then given by

$$T(h): \ (x, v, u_1, u_2) \mapsto \big(x, g(x)v, u_1, g'(x)u_1 \cdot v + g(x)u_2\big).$$

If we pull back the canonical 2-form described in Proposition 7.2 of Chapter V from $T^\vee(U) \approx T(U)$ to $T(U)$ by means of h then its description locally can be written on $U \times \mathbf{E}$ in the following manner.

(1)　　$\langle \Omega_{(x, v)}, (u_1, u_2) \times (w_1, w_2) \rangle =$

$$\langle u_1, g(x)w_2 \rangle - \langle u_2, g(x)w_1 \rangle - \langle g'(x)u_1 \cdot v, w_1 \rangle + \langle g'(x)w_1 \cdot v, u_1 \rangle.$$

From the simple formula giving our canonical 2-form on the cotangent bundle in Chapter V, we see at once that it is nonsingular on $T(U)$. Since h is a VB-isomorphism, it follows that the pull-back of this 2-form to the tangent bundle is also non-singular.

We shall now apply the results of the preceding section. To do so, we construct a 1-form on $T(X)$. Indeed, we have a function **(kinetic energy!)**

$$K: \ T(X) \to \mathbf{R}$$

given by $K(v) = \frac{1}{2}\langle v, v \rangle_x$ if v is in T_x. Then dK is a 1-form. By Proposition 6.1 of Chapter V, it corresponds to a vector field on $T(X)$, and we contend:

Theorem 7.1. *The vector field F on $T(X)$ corresponding to $-dK$ under the canonical 2-form is a spray over X, called the* **canonical spray.**

Proof. We work locally. We take U open in $\mathbf{E}$ and have the double tangent bundle

$$(U \times \mathbf{E}) \times (\mathbf{E} \times \mathbf{E})$$

$$\downarrow$$

$$U \times \mathbf{E}$$

$$\downarrow$$

$$U.$$

Our function K can be written

$$K(x, v) = \tfrac{1}{2}\langle v, v \rangle_x = \tfrac{1}{2}\langle v, g(x)v \rangle,$$

and dK at a point (x, v) is simply the ordinary derivative

$$DK(x, v)\colon \mathbf{E} \times \mathbf{E} \to \mathbf{R}.$$

The derivative DK is completely described by the two partial derivatives, and we have

$$DK(x, v) \cdot (w_1, w_2) = D_1 K(x, v) \cdot w_1 + D_2 K(x, v) \cdot w_2.$$

From the definition of derivative, we find

$$D_1 K(x, v) \cdot w_1 = \tfrac{1}{2}\langle v, g'(x)w_1 \cdot v \rangle$$
$$D_2 K(x, v) \cdot w_2 = \langle w_2, g(x)v \rangle = \langle v, g(x)w_2 \rangle.$$

We use the notation of Proposition 3.2 of Chapter IV. We can represent the vector field F corresponding to dK under the canonical 2-form Ω by a morphism $f\colon U \times \mathbf{E} \to \mathbf{E} \times \mathbf{E}$, which we write in terms of its two components:

$$f(x, v) = \big(f_1(x, v), f_2(x, v)\big) = (u_1, u_2).$$

Then by definition:

$$(2) \quad \langle \Omega_{(x,v)}, \big(f_1(x, v), f_2(x, v)\big) \times (w_1, w_2) \rangle = \langle DK(x, v), (w_1, w_2) \rangle$$
$$= D_1 K(x, v) \cdot w_1 + \langle v, g(x)w_2 \rangle.$$

Comparing expressions (1) to (2), we find that as functions of w_2 they have only one term on the right side depending on w_2. From the equality of the two expressions, we conclude that

$$\langle f_1(x, v), g(x)w_2 \rangle = \langle v, g(x)w_2 \rangle$$

for all w_2, and hence that $f_1(x, v) = v$, whence our vector field F is a second order vector field on X.

Again we compare expression (1) and (2), using the fact just proved that $u_1 = f_1(x, v) = v$. Setting the right sides of the two expressions equal to each other, and using $u_2 = f_2(u, v)$, we obtain:

Proposition 7.2. *In the chart U, let $f = (f_1, f_2)$: $U \times \mathbf{E} \to \mathbf{E} \times \mathbf{E}$ represent F. Then $f_2(x, v)$ is the unique vector such that for all $w_1 \in \mathbf{E}$ we have*:

$$\langle f_2(x, v), g(x)w_1 \rangle = \tfrac{1}{2} \langle g'(x)w_1 \cdot v, v \rangle - \langle g'(x) \cdot v \cdot v, w_1 \rangle.$$

From this one sees that f_2 is homogeneous of degree 2 in the second variable v, in other words that it represents a spray. This concludes the proof of Theorem 7.1.

CHAPTER VIII

Integration of Differential Forms -

The material of this chapter is also contained in my book on real analysis [La 93], but it may be useful to the reader to have it also here in a rather self contained way, based only on standard properties of integration in Euclidean space.

Throughout this chapter, μ is Lebesgue measure on $\mathbf{R}^n$.
If A is a subset of $\mathbf{R}^n$, we write $\mathscr{L}^1(A)$ instead of $\mathscr{L}^1(A, \mu, \mathbf{C})$.
Manifolds may have a boundary.

VIII, §1. SETS OF MEASURE 0

We recall that a set has **measure 0** in $\mathbf{R}^n$ if and only if, given ϵ, there exists a covering of the set by a sequence of rectangles $\{R_j\}$ such that $\sum \mu(R_j) < \epsilon$. We denote by R_j the closed rectangles, and we may always assume that the interiors R_j^0 cover the set, at the cost of increasing the lengths of the sides of our rectangles very slightly (an $\epsilon/2^n$ argument). We shall prove here some criteria for a set to have measure 0. We leave it to the reader to verify that instead of rectangles, we could have used cubes in our characterization of a set of a measure 0 (a cube being a rectangle all of whose sides have the same length).

We recall that a map f satisfies a **Lipschitz condition** on a set A if there exists a number C such that

$$|f(x) - f(y)| \leq C|x - y|$$

for all $x, y \in A$. Any C^1 map f satisfies locally at each point a Lipschitz condition, because its derivative is bounded in a neighborhood of each point, and we can then use the mean value estimate,

$$|f(x) - f(y)| \leqq |x - y| \sup|f'(z)|,$$

the sup being taken for z on the segment between x and y. We can take the neighborhood of the point to be a ball, say, so that the segment between any two points is contained in the neighborhood.

Lemma 1.1. *Let A have measure 0 in $\mathbf{R}^n$ and let $f: A \to \mathbf{R}^n$ satisfy a Lipschitz condition. Then $f(A)$ has measure 0.*

Proof. Let C be a Lipschitz constant for f. Let $\{R_j\}$ be a sequence of cubes covering A such that $\sum \mu(R_j) < \epsilon$. Let r_j be the length of the side of R_j. Then for each j we see that $f(A \cap S_j)$ is contained in a cube R_j' whose sides have length $\leqq 2Cr_j$. Hence

$$\mu(R_j') \leqq 2^n C^n r_j^n = 2^n C^n \mu(R_j).$$

Our lemma follows.

Lemma 1.2. *Let U be open in $\mathbf{R}^n$ and let $f: U \to \mathbf{R}^n$ be a C^1 map. Let Z be a set of measure 0 in U. Then $f(Z)$ has measure 0.*

Proof. For each $x \in U$ there exists a rectangle R_x contained in U such that the family $\{R_x^0\}$ of interiors covers Z. Since U is separable, there exists a denumerable subfamily covering Z, say $\{R_j\}$. It suffices to prove that $f(Z \cap R_j)$ has measure 0 for each j. But f satisfies a Lipschitz condition on R_j since R_j is compact and f' is bounded on R_j, being continuous. Our lemma follows from Lemma 1.1.

Lemma 1.3. *Let A be a subset of $\mathbf{R}^m$. Assume that $m < n$. Let*

$$f: A \to \mathbf{R}^n$$

satisfy a Lipschitz condition. Then $f(A)$ has measure 0.

Proof. We view $\mathbf{R}^m$ as embedded in $\mathbf{R}^n$ on the space of the first m coordinates. Then $\mathbf{R}^m$ has measure 0 in $\mathbf{R}^n$, so that A has also n-dimensional measure 0. Lemma 1.3 is therefore a consequence of Lemma 1.1.

Note. All three lemmas may be viewed as stating that certain parametrized sets have measure 0. Lemma 1.3 shows that parametrizing a set by strictly lower dimensional spaces always yields an image having measure 0. The other two lemmas deal with a map from one space into another of the same dimension. Observe that Lemma 1.3 would be false if f is only assumed to be continuous (Peano curves).

The next theorem will be used later only in the proof of the residue theorem, but it is worthwhile inserting it at this point.

Let $f: X \to Y$ be a morphism of class C^p, with $p \geq 1$, and assume throughout this section that X, Y are finite dimensional. A point $x \in X$ is called a **critical point** of f if f is not a submersion at x. This means that

$$T_x f: \ T_x X \to T_{f(x)} Y$$

is not surjective, according to our differrential criterion for a submersion.

Assume that a manifold X has a countable base for its charts. Then we can say that a set has measure 0 in X if its intersection with each chart has measure 0.

Theorem 1.4 (Sard's Theorem). *Let* $f: X \to Y$ *be a* C^∞ *morphism of manifolds having a countable base. Let* Z *be the set of critical points of* f *in* X. *Then* $f(Z)$ *has measure 0 in* Y.

Proof. (Due to Dieudonné.) By induction on the dimension n of X. The assertion is trivial if $n = 0$. Assume $n \geq 1$. It will suffice to prove the theorem locally in the neighborhood of a point in Z. We may assume that $X = U$ is open in $\mathbf{R}^n$ and

$$f: \ U \to \mathbf{R}^p$$

can be expressed in terms of coordinate functions,

$$f = (f_1, \ldots, f_p).$$

We let us usual

$$D^\alpha = D_1^{\alpha_1} \cdots D_n^{\alpha_n}$$

be a differential operator, and call $|\alpha| = \alpha_1 + \cdots + \alpha_n$ its **order**. We let $Z_0 = Z$ and for $m \geq 1$ we let Z_m be the set of points $x \in Z$ such that

$$D^\alpha f_j(x) = 0$$

for all $j = 1, \ldots, p$ and all α with $1 \leq |\alpha| \leq m$. We shall prove:

(1) *For each* $m \geq 0$ *the set* $f(Z_m - Z_{m+1})$ *has measure 0.*

(2) *If* $m \geq n/p$, *then* $f(Z_m)$ *has measure 0.*

This will obviously prove Sard's theorem.

Proof of (1). Let $a \in Z_m - Z_{m+1}$. Suppose first that $m = 0$. Then for some coordinate function, say $j = 1$, and after a renumbering of the variables if necessary, we have

$$D_1 f_1(a) \neq 0.$$

The map

$$g: \ x \mapsto (f_1(x), x_2, \ldots, x_p)$$

obviously has an invertible derivative at $x = a$, and hence is a local isomorphism at a. Considering $f \circ g^{-1}$ instead of f, we are reduced to the case where f is given by

$$f(x) = (x_1, f_2(x), \ldots, f_p(x)) = (x_1, h(x)),$$

where h is the projection of f on the last $p - 1$ coordinates and is therefore a morphism $h: \ V \to \mathbf{R}^{p-1}$ defined on some open V containing a. Then

$$Df(x) = \begin{pmatrix} 1 & 0 \\ * & Dh(x) \end{pmatrix}.$$

From this it is clear that x is a critical point for f if and only if x is a critical point for h, and it follows that $h(Z \cap V)$ has measure 0 in $\mathbf{R}^{p-1}$. Since $f(Z)$ is contained in $\mathbf{R}^1 \times h(Z)$, we conclude that $f(Z)$ has measure 0 in $\mathbf{R}^p$ as desired.

Next suppose that $m \geq 1$. Then for some α with $|\alpha| = m + 1$, and say $j = 1$, we have

$$D^\alpha f_1(a) \neq 0.$$

Again after a renumbering of the indices, we may write

$$D^\alpha f_1 = D_1 g_1$$

for some function g_1, and we observe that $g_1(x) = 0$ for all $x \in Z_m$, in a neighborhood of a. The map

$$g: \ x \mapsto (g_1(x), x_2, \ldots, x_n)$$

is then a local isomorphism at a, say on an open set V containing a, and we see that

$$g(Z_m \cap V) \subset \{0\} \times \mathbf{R}^{n-1}.$$

We view g as a change of charts, and considering $f \circ g^{-1}$ instead of f, together with the invariance of critical points under changes of charts, we may view f as defined on an open subset of $\mathbf{R}^{n-1}$. We can then apply induction again to conclude the proof of our first assertion.

Proof of (2). Again we work locally, and we may view f as defined on the closed n-cube of radius r centered at some point a. We denote this cube by $C_r(a)$. For $m \geq n/p$, it will suffice to prove that

$$f\left(Z_m \cap C_r(a)\right)$$

has measure 0. For large N, we cut up each side of the cube into N equal segments, thus obtaining a decomposition of the cube into N^n small cubes. By Taylor's formula, if a small cube contains a critical point $x \in Z_m$, then for any point y of this small cube we have

$$|f(y) - f(x)| \leq K|x - y|^{m+1} \leq K(2r/N)^{m+1},$$

where K is a bound for the derivatives of f up to order $m + 1$, and we use the sup norm. Hence the image of Z_m contained in small cube is itself contained in a cube whose radius is given by the right-hand side, and whose volume in $\mathbf{R}^p$ is therefore bounded by

$$K^p(2r/N)^{p(m+1)}.$$

We have at most N^n such images to consider and we therefore see that

$$f\left(Z_m \cap C_r(a)\right)$$

is contained in a union of cubes in $\mathbf{R}^p$, the sum of whose volumes is bounded by

$$K^p N^n (2r/N)^{p(m+1)} \leq K^p (2r)^{p(m+1)} N^{n-p(m+1)}.$$

Since $m \geq n/p$, we see that the right-hand side of this estimate behaves like $1/N$ as N becomes large, and hence that the union of the cubes in $\mathbf{R}^p$ has arbitrarily small measure, thereby proving Sard's theorem.

Sard's theorem is harder to prove in the case f is C^p with finite p, see [Str 64/83], but $p = \infty$ already is quite useful.

VIII, §2. CHANGE OF VARIABLES FORMULA

We first deal with the simplest of cases. We consider vectors $v_1, \ldots, v_n$ in $\mathbf{R}^n$ and we define the **block** B spanned by these vectors to be the set of points

$$t_1 v_1 + \cdots + t_n v_n$$

with $0 \leq t_i \leq 1$. We say that the block is **degenerate** (in $\mathbf{R}^n$) if the vectors

$v_1, \ldots, v_n$ are linearly dependent. Otherwise, we say that the block is **non-degenerate**, or is a **proper block** in $\mathbf{R}^n$.

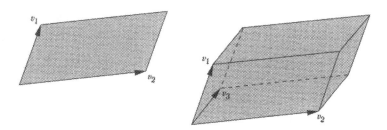

We see that a block in $\mathbf{R}^2$ is nothing but a parallelogram, and a block in $\mathbf{R}^3$ is nothing but a parallelepiped (when not degenerate).

We shall sometimes use the word volume instead of measure when applied to blocks or their images under maps, for the sake of geometry.

We denote by $\mathrm{Vol}(v_1, \ldots, v_n)$ the volume of the block B spanned by $v_1, \ldots, v_n$. We define the **oriented volume**

$$\mathrm{Vol}^0(v_1, \ldots, v_n) = \pm \mathrm{Vol}(v_1, \ldots, v_n),$$

taking the $+$ sign if $\mathrm{Det}(v_1, \ldots, v_n) > 0$ and the $-$ sign if

$$\mathrm{Det}(v_1, \ldots, v_n) < 0.$$

The determinant is viewed as the determinant of the matrix whose column vectors are $v_1, \ldots, v_n$, in that order.

We recall the following characterization of determinants. Suppose that we have a product

$$(v_1, \ldots, v_n) \mapsto v_1 \wedge v_2 \wedge \cdots \wedge v_n$$

which to each n-tuple of **vectors** associates a number, such that the product is multilinear, alternating, and such that

$$e_1 \wedge \cdots \wedge e_n = 1$$

if $e_1, \ldots, e_n$ are the unit vectors. Then this product is necessarily the determinant, that is, it is uniquely determined. "Alternating" means that if $v_i = v_j$ for some $i \neq j$, then

$$v_1 \wedge \cdots \wedge v_n = 0.$$

The uniqueness is easily proved, and we recall this short proof. We can write

$$v_i = a_{i1}e_1 + \cdots + a_{in}e_n$$

for suitable numbers a_{ij}, and then

$$v_1 \wedge \cdots \wedge v_n = (a_{11}e_1 + \cdots + a_{1n}e_n) \wedge \cdots \wedge (a_{n1}e_1 + \cdots + a_{nn}e_n)$$

$$= \sum_\sigma a_{1,\sigma(1)}e_{\sigma(1)} \wedge \cdots \wedge a_{n,\sigma(n)}e_{\sigma(n)}$$

$$= \sum_\sigma a_{1,\sigma(1)} \cdots a_{n,\sigma(n)}e_{\sigma(1)} \wedge \cdots \wedge e_{\sigma(n)}.$$

The sum is taken over all maps $\sigma\colon \{1,\ldots,n\} \to \{1,\ldots,n\}$, but because of the alternating property, whenever σ is not a permutation the term corresponding to σ is equal to 0. Hence the sum may be taken only over all permutations. Since

$$e_{\sigma(1)} \wedge \cdots \wedge e_{\sigma(n)} = \epsilon(\sigma)e_1 \wedge \cdots \wedge e_n$$

where $\epsilon(\sigma) = 1$ or -1 is a sign depending only on σ, it follows that the alternating product is completely determined by its value $e_1 \wedge \cdots \wedge e_n$, and in particular is the determinant if this value is equal to 1.

Proposition 2.1. *We have*

$$\mathrm{Vol}^0(v_1,\ldots,v_n) = \mathrm{Det}(v_1,\ldots,v_n)$$

and

$$\mathrm{vol}(v_1,\ldots,v_n) = |\mathrm{Det}(v_1,\ldots,v_n)|.$$

Proof. If $v_1,\ldots,v_n$ are linearly dependent, then the determinant is equal to 0, and the volume is also equal to 0, for instance by Lemma 1.3. So our formula holds in the case. It is clear that

$$\mathrm{Vol}^0(e_1,\ldots,e_n) = 1.$$

To show that Vol^0 satisfies the characteristic properties of the determinant, all we have to do now is to show that it is linear in each variable, say the first. In other words, we must prove

$(*)$ $\mathrm{Vol}^0(cv, v_2,\ldots,v_n) = c\,\mathrm{Vol}^0(v, v_2,\ldots,v_n)$ for $c \in \mathbf{R}$,

$(**)$ $\mathrm{Vol}^0(v + w, v_2,\ldots,v_n) = \mathrm{Vol}^0(v, v_2,\ldots,v_n) + \mathrm{Vol}^0(w, v_2,\ldots,v_n).$,

As to the first assertion, suppose first that c is some positive integer k. Let B be the block spanned by $v, v_2,\ldots,v_n$. We may assume without loss of generality that $v, v_2,\ldots,v_n$ are linearly independent (otherwise, the relation is obviously true, both sides being equal to 0). We verify at once from the definition that if $B(v, v_2,\ldots,v_n)$ denotes the block spanned by $v, v_2,\ldots,v_n$

then $B(kv, v_2, \ldots, v_n)$ is the union of the two sets

$$B\big((k-1)v, v_2, \ldots, v_n\big) \qquad \text{and} \qquad B(v, v_2, \ldots, v_n) + (k-1)v$$

which have only a set of measure 0 in common, as one verifies at once from the definitions.

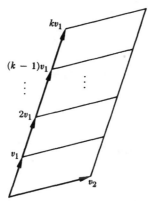

Therefore, we find that

$$\begin{aligned}
\mathrm{Vol}(kv, v_2, \ldots, v_n) &= \mathrm{Vol}\big((k-1)v, v_2, \ldots, v_n\big) + \mathrm{Vol}(v, v_2, \ldots, v_n) \\
&= (k-1)\,\mathrm{Vol}(v, v_2, \ldots, v_n) + \mathrm{Vol}(v, v_2, \ldots, v_n) \\
&= k\,\mathrm{Vol}(v, v_2, \ldots, v_n),
\end{aligned}$$

as was to be shown.

Now let

$$v = v_1/k$$

for a positive integer k. Then applying what we have just proved shows that

$$\mathrm{Vol}\left(\frac{1}{k}v_1, v_2, \ldots, v_n\right) = \frac{1}{k}\,\mathrm{Vol}(v_1, \ldots, v_n).$$

Writing a positive rational number in the form $m/k = m \cdot 1/k$, we conclude that the first relation holds when c is a positive rational number. If r is a positive real number, we find positive rational numbers c, c' such that $c \leqq r \leqq c'$. Since

$$B(cv, v_2, \ldots, v_n) \subset B(rv, v_2, \ldots, v_n) \subset B(c'v, v_2, \ldots, v_n),$$

we conclude that

$$c\,\mathrm{Vol}(v, v_2, \ldots, v_n) \leqq \mathrm{Vol}(rv, v_2, \ldots, v_n) \leqq c'\,\mathrm{Vol}(v, v_2, \ldots, v_n).$$

Letting c, c' approach r as a limit, we conclude that for any real number $r \geq 0$ we have

$$\mathrm{Vol}(rv, v_2, \ldots, v_n) = r\,\mathrm{Vol}(v, v_2, \ldots, v_n).$$

Finally, we note that $B(-v, v_2, \ldots, v_n)$ is the translation of

$$B(v, v_2, \ldots, v_n)$$

by $-v$ so that these two blocks have the same volume. This proves the first assertion.

As for the second, we look at the geometry of the situation, which is made clear by the following picture in case $v = v_1$, $w = v_2$.

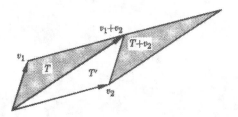

The block spanned by $v_1, v_2, \ldots$ consists of two "triangles" T, T' having only a set of measure zero in common. The block spanned by $v_1 + v_2$ and v_2 consists of T' and the translation $T + v_2$. It follows that these two blocks have the same volume. We conclude that for any number c,

$$\mathrm{Vol}^0(v_1 + cv_2, v_2, \ldots, v_n) = \mathrm{Vol}^0(v_1, v_2, \ldots, v_n).$$

Indeed, if $c = 0$ this is obvious, and if $c \neq 0$ then

$$c\,\mathrm{Vol}^0(v_1 + cv_2, v_2) = \mathrm{Vol}^0(v_1 + cv_2, cv_2)$$
$$= \mathrm{Vol}^0(v_1 + cv_2) = c\,\mathrm{Vol}^0(v_1, v_2).$$

We can then cancel c to get our conclusion.

To prove the linearity of Vol^0 with respect to its first variable, we may assume that $v_2, \ldots, v_n$ are linearly independent, otherwise both sides of (**) are equal to 0. Let v_1 be so chosen that $\{v_1, \ldots, v_n\}$ is a basis of $\mathbf{R}^n$. Then by induction, and what has been proved above,

$$\mathrm{Vol}^0(c_1 v_1 + \cdots + c_n v_n, v_2, \ldots, v_n)$$
$$= \mathrm{Vol}^0(c_1 v_1 + \cdots + c_{n-1} v_{n-1}, v_2, \ldots, v_n)$$
$$= \mathrm{Vol}^0(c_1 v_1, v_2, \ldots, v_n)$$
$$= c_1 \mathrm{Vol}^0(v_1, \ldots, v_n).$$

From this the linearity follows at once, and the theorem is proved.

Corollary 2.2. *Let S be the unit cube spanned by the unit vectors in $\mathbf{R}^n$. Let $\lambda\colon \mathbf{R}^n \to \mathbf{R}^n$ be a linear map. Then*

$$\operatorname{Vol} \lambda(S) = |\operatorname{Det}(\lambda)|.$$

Proof. If $v_1, \ldots, v_n$ are the images of $e_1, \ldots, e_n$ under λ, then $\lambda(S)$ is the block spanned by $v_1, \ldots, v_n$. If we represent λ by the matrix $A = (a_{ij})$, then

$$v_i = a_{1i}e_1 + \cdots + a_{ni}e_n,$$

and hence $\operatorname{Det}(v_1, \ldots, v_n) = \operatorname{Det}(A) = \operatorname{Det}(\lambda)$. This proves the corollary.

Corollary 2.3. *If R is any rectangle in $\mathbf{R}^n$ and $\lambda\colon \mathbf{R}^n \to \mathbf{R}^n$ is a linear map, then*
$$\operatorname{Vol} \lambda(R) = |\operatorname{Det}(\lambda)|\operatorname{Vol}(R).$$

Proof. After a translation, we can assume that the rectangle is a block. If $R = \lambda_1(S)$ where S is the unit cube, then

$$\lambda(R) = \lambda \circ \lambda_1(S),$$

whence by Corollary 2.2,

$$\operatorname{Vol} \lambda(R) = |\operatorname{Det}(\lambda \circ \lambda_1)| = |\operatorname{Det}(\lambda)\,\operatorname{Det}(\lambda_1)| = |\operatorname{Det}(\lambda)|\,\operatorname{Vol}(R).$$

The next theorem extends Corollary 2.3 to the more general case where the linear map λ is replaced by an arbitrary C^1-invertible map. The proof then consists of replacing the linear map by its derivative and estimating the error thus introduced. For this purpose, we have the **Jacobian determinant**

$$\Delta_f(x) = \operatorname{Det} J_f(x) = \operatorname{Det} f'(x),$$

where $J_f(x)$ is the Jacobian matrix, and $f'(x)$ is the derivative of the map $f\colon U \to \mathbf{R}^n$.

Proposition 2.4. *Let R be a rectangle in $\mathbf{R}^n$, contained in some open set U. Let $f\colon U \to \mathbf{R}^n$ be a C^1 map, which is C^1-invertible on U. Then*

$$\mu\big(f(R)\big) = \int_R |\Delta_f|\, d\mu.$$

Proof. When f is linear, this is nothing but Corollary 2.3 of the preceding theorem. We shall prove the general case by approximating f by its derivative. Let us first assume that R is a cube for simplicity. Given ϵ, let P be a partition of R, obtained by dividing each side of R into N

equal segments for large N. Then R is partitioned into N^n subcubes which we denote by S_j $(j = 1, \ldots, N^n)$. We let a_j be the center of S_j.

We have

$$\text{Vol } f(R) = \sum_j \text{Vol } f(S_j)$$

because the images $f(S_j)$ have only sets of measure 0 in common. We investigate $f(S_j)$ for each j. The derivative f' is uniformly continuous on R. Given ϵ, we assume that N has been taken so large that for $x \in S_j$ we have

$$f(x) = f(a_j) + \lambda_j(x - a_j) + \varphi(x - a_j),$$

where $\lambda_j = f'(a_j)$ and

$$|\varphi(x - a_j)| \leqq |x - a_j|\epsilon.$$

To determine $\text{Vol } f(S_j)$ we must therefore investigate $f(S)$ where S is a cube centered at the origin, and f has the form

$$f(x) = \lambda x + \varphi(x), \qquad |\varphi(x)| \leqq |x|\epsilon.$$

on the cube S. (We have made suitable translations which don't affect volumes.) We have

$$\lambda^{-1} \circ f(x) = x + \lambda^{-1} \circ \varphi(x),$$

so that $\lambda^{-1} \circ f$ is nearly the identity map. For some constant C, we have for $x \in S$

$$|\lambda^{-1} \circ \varphi(x)| \leqq C\epsilon.$$

From the lemma after the proof of the inverse mapping theorem, we conclude that $\lambda^{-1} \circ f(S)$ contains a cube of radius

$$(1 - C\epsilon)(\text{radius } S),$$

and trivial estimates show that $\lambda^{-1} \circ f(S)$ is contained in a cube of radius

$$(1 + C\epsilon)(\text{radius } S).$$

We apply λ to these cubes, and determine their volumes. Putting indices j on everything, we find that

$$|\text{Det } f'(a_j)| \, \text{Vol}(S_j) - \epsilon C_1 \text{Vol}(S_j)$$
$$\leqq \text{Vol } f(S_j) \leqq |\text{Det } f'(a_j)| \, \text{Vol}(S_j) + \epsilon C_1 \text{Vol}(S_j)$$

with some fixed constant C_1. Summing over j and estimating $|\Delta_f|$, we see that our theorem follows at once.

Remark. We assumed for simplicity that R was a cube. Actually, by changing the norm on each side, multiplying by a suitable constant, and taking the sup of the adjusting norms, we see that this involves no loss of generality. Alternatively, we can approximate a given rectangle by cubes.

Corollary 2.5. *If g is continuous on $f(R)$, then*

$$\int_{f(R)} g \, d\mu = \int_R (g \circ f)|\Delta_f| \, d\mu.$$

Proof. The functions g and $(g \circ f)|\Delta_f|$ are uniformly continuous on $f(R)$ and R respectively. Let us take a partition of R and let $\{S_j\}$ be the subrectangles of this partition. If δ is the maximum length of the sides of the subrectangles of the partition, then $f(S_j)$ is contained in a rectangle whose sides have length $\leq C\delta$ for some constant C. We have

$$\int_{f(R)} g \, d\mu = \sum_j \int_{f(S_j)} g \, d\mu.$$

The sup and inf of g of $f(S_j)$ differ only by ϵ if δ is taken sufficiently small. Using the theorem, applied to each S_j, and replacing g by its minimum m_j and maximum M_j on S_j, we see that the corollary follows at once.

Theorem 2.6 (Change of Variables Formula). *Let U be open in $\mathbf{R}^n$ and let $f \colon U \to \mathbf{R}^n$ be a C^1 map, which is C^1 invertible on U. Let g be in $\mathscr{L}^1(f(U))$. Then $(g \circ f)|\Delta_f|$ is in $\mathscr{L}^1(U)$ and we have*

$$\int_{f(U)} g \, d\mu = \int_U (g \circ f)|\Delta_f| \, d\mu.$$

Proof. Let R be a closed rectangle contained in U. We shall first prove that the restriction of $(g \circ f)|\Delta_f|$ to R is in $\mathscr{L}^1(R)$, and that the formula holds when U is replaced by R. We know that $C_c(f(U))$ is L^1-dense in $\mathscr{L}^1(f(U))$, by [La 93], Theorem 3.1 of Chapter IX. Hence there exists a sequence $\{g_k\}$ in $C_c(f(U))$ which in L^1-convergent to g. Using [La 93], Theorem 5.2 of Chapter VI, we may assume that $\{g_k\}$ converges pointwise to g except on a set Z of measure 0 in $f(U)$. By Lemma 1.2, we know that $f^{-1}(Z)$ has measure 0.

Let $g_k^* = (g_k \circ f)|\Delta_f|$. Each function g_k^* is continuous on R. The sequence $\{g_k^*\}$ converges almost everywhere to $(g \circ f)|\Delta_f|$ restricted to R.

It is in fact an L^1-Cauchy sequence in $\mathcal{L}^1(R)$. To see this, we have by the result for rectangles and continuous functions (corollary of the preceding theorem):

$$\int_R |g_k^* - g_m^*| \, d\mu = \int_{f(R)} |g_k - g_m| \, d\mu,$$

so the Cauchy nature of the sequence $\{g_k^*\}$ is clear from that of $\{g_k\}$. It follows that the restriction of $(g \circ f)|\Delta_f|$ to R is the L^1-limit of $\{g_k^*\}$, and is in $\mathcal{L}^1(R)$. It also follows that the formula of the theorem holds for R, that is

$$\int_{f(A)} g \, d\mu = \int_A (g \circ f)|\Delta_f| \, d\mu$$

when $A = R$.

The theorem is now seen to hold for any measurable subset A of R, since $f(A)$ is measurable, and since a function g in $\mathcal{L}^1(f(A))$ can be extended to a function in $\mathcal{L}^1(f(R))$ by giving it the value 0 outside $f(A)$. From this it follows that the theorem holds if A is a finite union of rectangles contained in U. We can find a sequence of rectangles $\{R_m\}$ contained in U whose union is equal to U, because U is separable. Taking the usual stepwise complementation, we can find a disjoint sequence of measurable sets

$$A_m = R_m - (R_1 \cup \cdots \cup R_{m-1})$$

whose union is U, and such that our theorem holds if $A = A_m$. Let

$$h_m = g_{f(A_m)} = g\chi_{f(A_m)} \quad \text{and} \quad h_m^* = (h_m \circ f)|\Delta_f|.$$

Then $\sum h_m$ converges to g and $\sum h_m^*$ converges to $(g \circ f)|\Delta_f|$. Our theorem follows from Corollary 5.13 of the dominated convergence theorem in [La 93].

Note. In dealing with polar coordinates or the like, one sometimes meets a map f which is invertible except on a set of measure 0, e.g. the polar coordinate map. It is now trivial to recover a result covering this type of situation.

Corollary 2.7. *Let U be open in $\mathbf{R}^n$ and let $f: U \to \mathbf{R}^n$ be a C^1 map. Let A be a measurable subset of U such that the boundary of A has measure 0, and such that f is C^1 invertible on the interior of A. Let g be in $\mathcal{L}^1(f(A))$. Then $(g \circ f)|\Delta_f|$ is in $\mathcal{L}^1(A)$ and*

$$\int_{f(A)} g \, d\mu = \int_A (g \circ f)|\Delta_f| \, d\mu.$$

Proof. Let U_0 be the interior of A. The sets $f(A)$ and $f(U_0)$ differ only by a set of measure 0, namely $f(\partial A)$. Also the sets A, U_0 differ only by a set of measure 0. Consequently we can replace the domains of integration $f(A)$ and A by $f(U_0)$ and U_0, respectively. The theorem applies to conclude the proof of the corollary.

VIII, §3. ORIENTATION

Let U, V be open sets in half spaces of $\mathbf{R}^n$ and let $\varphi\colon U \to V$ be a C^1 isomorphism. We shall say that φ is **orientation preserving** if the Jacobian determinant $\Delta_\varphi(x)$ is > 0, all $x \in U$. If the Jacobian determinant is negative, then we say that φ is orientation **reversing**.

Let X be a C^p manifold, $p \geq 1$, and let $\{(U_i, \varphi_i)\}$ be an atlas. We say that this atlas is **oriented** if all transition maps $\varphi_j \circ \varphi_i^{-1}$ are orientation preserving. Two atlases $\{(U_i, \varphi_i)\}$ and $\{(V_\alpha, \psi_\alpha)\}$ are said to **define the same orientation**, or to be **orientation equivalent**, if their union is oriented. We can also define locally a chart (V, ψ) to be **orientation compatible** with the oriented atlas $\{(U_i, \varphi_i)\}$ if all transition maps $\varphi_i \circ \varphi^{-1}$ (defined whenever $U_i \cap V$ is not empty) are orientation preserving. An orientation equivalence class of oriented atlases is said to define an **oriented** manifold, or to be an **orientation** of the manifold. It is a simple exercise to verify that if a connected manifold has an orientation, then it has two distinct orientations.

The standard examples of the Moebius strip or projective plane show that not all manifolds admit orientations. We shall now see that the boundary of an oriented manifold with boundary can be given a natural orientation.

Let $\varphi\colon U \to \mathbf{R}^n$ be an oriented chart at a boundary point of X, such that:

(1) *if $(x_1, \ldots, x_n)$ are the local coordinates of the chart, then the boundary points correspond to those points in $\mathbf{R}^n$ satisfying $x_1 = 0$; and*

(2) *the points of U not in the boundary have coordinates satisfying $x_1 < 0$.*

Then $(x_2, \ldots, x_n)$ are the local coordinates for a chart of the boundary, namely the restriction of φ to $\partial X \cap U$, and the picture is as follows.

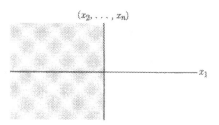

We may say that we have considered a chart φ such that the manifold lies to the left of its boundary. If the reader thinks of a domain in $\mathbf{R}^2$, having a smooth curve for its boundary, as on the following picture, the reader will see that our choice of chart corresponds to what is usually visualized as "counterclockwise" orientation.

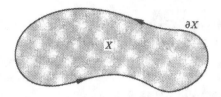

The collection of all pairs $\bigl(U \cap \partial X, \varphi|(U \cap \partial X)\bigr)$, chosen according to the criteria described above, is obviously an atlas for the boundary ∂X, and we contend that it is an oriented atlas.

We prove this easily as follows. If

$$(x_1, \ldots, x_n) = x \quad \text{and} \quad (y_1, \ldots, y_n) = y$$

are coordinate systems at a boundary point corresponding to choices of charts made according to our specifications, then we can write $y = f(x)$ where $f = (f_1, \ldots, f_n)$ is the transition mapping. Since we deal with oriented charts for X, we know that $\Delta_f(x) > 0$ for all x. Since f maps boundary into boundary, we have

$$f_1(0, x_2, \ldots, x_n) = 0$$

for all $x_2, \ldots, x_n$. Consequently the Jacobian matrix of f at a point $(0, x_2, \ldots, x_n)$ is equal to

$$\begin{pmatrix} D_1 f_1(0, x_2, \ldots, x_n) & 0 \cdots \cdots 0 \\ * & \\ * & \Delta_g^{(n-1)} \\ * & \end{pmatrix},$$

where $\Delta_g^{(n-1)}$ is the Jacobian matrix of the transition map g induced by f on the boundary, and given by

$$\begin{aligned} y_2 &= f_2(0, x_2, \ldots, x_n), \\ &\vdots \qquad \vdots \\ y_n &= f_n(0, x_2, \ldots, x_n). \end{aligned}$$

However, we have

$$D_1 f_1(0, x_2, \ldots, x_n) = \lim_{h \to 0} \frac{f_1(h, x_2, \ldots, x_n)}{h},$$

taking the limit with $h < 0$ since by prescription, points of X have coordinates with $x_1 < 0$. Furthermore, for the same reason we have

$$f_1(h, x_2, \ldots, x_n) < 0.$$

Consequently

$$D_1 f_1(0, x_2, \ldots, x_n) > 0.$$

From this it follows that $\Delta_g^{(n-1)}(x_2, \ldots, x_n) > 0$, thus proving our assertion that the atlas we have defined for ∂X is oriented.

From now on, when we deal with an oriented manifold, it is understood that its boundary is taken with orientation described above, and called the induced orientation.

VIII, §4. THE MEASURE ASSOCIATED WITH A DIFFERENTIAL FORM

Let X be a manifold of class C^p with $p \geq 1$. We assume from now on that X has a countable base. Then we know that X admits C^p partitions of unity, subordinated to any given open covering.

(Actually, instead of the conditions we assumed, we could just as well have assumed the existence of C^p partitions of unity, which is the precise condition to be used in the sequel.)

We can define the **support** of a differential form as we defined the support of a function. It is the closure of the set of all $x \in X$ such that $\omega(x) \neq 0$. If ω is a form of class C^p and α is a C^q function on X, then we can form the product $\alpha\omega$, which is the form whose value at x is $\alpha(x)\omega(x)$. If α has compact support, then $\alpha\omega$ has compact support. Later, we shall study the integration of forms, and reduce this to a local problem by means of partitions of unity, in which we multiply a form by functions.

We assume that the reader is familiar with the correspondence between certain functionals on continuous functions with compact support and measures. Cf. [La 93] for this. We just recall some terminology.

We denote by $C_c(X)$ the infinite dimensional vector space of continuous functions on X with **compact support** (i.e. vanishing outside a compact set). We write $C_c(X, \mathbf{R})$ or $C_c(X, \mathbf{C})$ if we wish to distinguish between the real or complex valued functions.

We denote by $C_K(X)$ the subspace of $C_c(X)$ consisting of those functions which vanish outside K. (Same notation $C_S(X)$ for those functions which are 0 outside any subset S of X. Most of the time, the useful subsets in this context are the compact subsets K.)

A linear map λ of $C_c(X)$ into the complex numbers (or into a normed vector space, for that matter) is said to be **bounded** if there exists some $C \geq 0$ such that we have for the sup norm

$$|\lambda f| \leq C\|f\|$$

for all $f \in C_c(X)$. Thus λ is bounded if and only if λ is continuous for the norm topology.

A linear map λ of $C_c(X)$ into the complex numbers is said to be **positive** if we have $\lambda f \geq 0$ whenever f is real and ≥ 0.

Lemma 4.1. *Let* $\lambda: C_c(X) \to \mathbf{C}$ *be a positive linear map. Then* λ *is bounded on* $C_K(X)$ *for any compact* K.

Proof. By the corollary of Urysohn's lemma, there exists a continuous real function $g \geq 0$ on X which is 1 on K has compact support. If $f \in C_K(X)$, let $b = \|f\|$. Say f is real. Then $bg \pm f \geq 0$, whence

$$\lambda(bg) \pm \lambda f \geq 0$$

and $|\lambda f| \leq b\lambda(g)$. Thus λg is our desired bound.

A complex valued linear map on $C_c(X)$ which is bounded on each subspace $C_K(X)$ for every compact K will be called a C_c-**functional** on $C_c(X)$, or more simply, a **functional**. A functional on $C_c(X)$ which is also continuous for the sup norm will be called a **bounded** functional. It is clear that a bounded functional is also a C_c-functional.

Lemma 4.2. *Let* $\{W_\alpha\}$ *be an open covering of* X. *For each index* α, *let* λ_α *be a functional on* $C_c(W_\alpha)$. *Assume that for each pair of indices* α, β *the functionals* λ_α *and* λ_β *are equal on* $C_c(W_\alpha \cap W_\beta)$. *Then there exists a unique functional* λ *on* X *whose restriction to each* $C_c(W_\alpha)$ *is equal to* λ_α. *If each* λ_α *is positive, then so is* λ.

Proof. Let $f \in C_c(X)$ and let K be the support of f. Let $\{h_i\}$ be a partition of unity over K subordinated to a covering of K by a finite number of the open sets W_α. Then each $h_i f$ has support in some $W_{\alpha(i)}$ and we define

$$\lambda f = \sum_i \lambda_{\alpha(i)}(h_i f).$$

We contend that this sum is independent of the choice of $\alpha(i)$, and also of the choice of partition of unity. Once this is proved, it is then obvious that λ is a functional which satisfies our requirements. We now prove this independence. First note that if $W_{\alpha'(i)}$ is another one of the open sets W_α in which the support of $h_i f$ is contained, then $h_i f$ has support in the intersection $W_{\alpha(i)} \cap W_{\alpha'(i)}$, and our assumption concerning our functionals λ_α shows that the corresponding term in the sum does not depend on the choice of index $\alpha(i)$. Next, let $\{g_k\}$ be another partition of unity over K subordinated to some covering of K by a finite number of the open sets W_α. Then for each i,

$$h_i f = \sum_k g_k\, h_i f,$$

whence

$$\sum_i \lambda_{\alpha(i)}(h_i f) = \sum_i \sum_k \lambda_{\alpha(i)}(g_k\, h_i f).$$

If the support of $g_k h_i f$ is in some W_α, then the value $\lambda_\alpha(g_k h_i f)$ is independent of the choice of index α. The expression on the right is then symmetric with respect to our two partitions of unity, whence our theorem follows.

Theorem 4.3. *Let* $\dim X = n$ *and let* ω *be an n-form on X of class* C^0, *that is continuous. Then there exists a unique positive functional* λ *on* $C_c(X)$ *having the following property. If* (U, φ) *is a chart and*

$$\omega(x) = f(x)\, dx_1 \wedge \cdots \wedge dx_n$$

is the local representation of ω *in this chart, then for any* $g \in C_c(X)$ *with support in* U, *we have*

(1)
$$\lambda g = \int_{\varphi U} g_\varphi(x) |f(x)|\, dx,$$

where g_φ *represents* g *in the chart* $[i.e.\ g_\varphi(x) = g(\varphi^{-1}(x))]$, *and* dx *is Lebesgue measure.*

Proof. The integral in (1) defines a positive functional on $C_c(U)$. The change of variables formula shows that if (U, φ) and (V, ψ) are two charts, and if g has support in $U \cap V$, then the value of the functional is independent of the choice of charts. Thus we get a positive functional by the general localization lemma for functionals.

The positive measure corresponding to the functional in Theorem 4.3 will be called the **measure associated with** $|\omega|$, and can be denoted by $\mu_{|\omega|}$.

Theorem 4.3 does not need any orientability assumption. With such an assumption, we have a similar theorem, obtained without taking the absolute value.

Theorem 4.4. *Let* dim $X = n$ *and assume that* X *is oriented. Let* ω *be an n-form on* X *of class* C^0. *Then there exists a unique functional* λ *on* $C_c(X)$ *having the following property. If* (U, φ) *is an oriented chart and*

$$\omega(x) = f(x) \, dx_1, \wedge \cdots \wedge dx_n$$

is the local representation of ω *in this chart, then for any* $g \in C_c(X)$ *with support in* U, *we have*

$$\lambda g = \int_{\varphi U} g_\varphi(x) f(x) \, dx,$$

where g_φ *represents* g *in the chart, and* dx *is Lebesgue measure.*

Proof. Since the Jacobian determinant of transition maps belonging to oriented charts is positive, we see that Theorem 4.4 follows like Theorem 4.3 from the change of variables formula (in which the absolute value sign now becomes unnecessary) and the existence of partitions of unity.

If λ is the functional of Theorem 4.4, we shall call it the **functional associated with** ω. For any function $g \in C_c(X)$, we define

$$\int_X g\omega = \lambda g.$$

If in particular ω has compact support, we can also proceed directly as follows. Let $\{\alpha_i\}$ be a partition of unity over X such that each α_i has compact support. We define

$$\int_X \omega = \sum_i \int_X \alpha_i \omega,$$

all but a finite number of terms in this sum being equal to 0. As usual, it is immediately verified that this sum is in fact independent of the choice of partition of unity, and in fact, we could just as well use only a partition of unity over the support of ω. Alternatively, if α is a function in $C_c(X)$ which is equal to 1 on the support of ω, then we could also define

$$\int_X \omega = \int_X \alpha\omega.$$

It is clear that these two possible definitions are equivalent. In particular, we obtain the following variation of Theorem 4.4.

Theorem 4.5. *Let X be an oriented manifold of dimension n. Let $\mathscr{A}_c^n(X)$ be the **R**-space of differential forms with compact support. There exists a unique linear map*

$$\omega \mapsto \int_X \omega \qquad of \quad \mathscr{A}_c^n(X) \to \mathbf{R}$$

such that, if ω has support in an oriented chart U with coordinates $x_1, \ldots, x_n$ and $\omega(x) = f(x)\, dx_1 \wedge \cdots \wedge dx_n$ in this chart, then

$$\int_X \omega = \int_U f(x)\, dx_1 \cdots dx_n.$$

Let X be an oriented manifold. By **a volume form** Ω we mean a form such that in every oriented chart, the form can be written as

$$\Omega(x) = f(x)\, dx_1 \wedge \cdots \wedge dx_n$$

with $f(x) > 0$ for all x. In Chapter X, §2 we shall see how to get a volume form from a Riemannian metric. For densities, see [La99] Chapter XVI, §4. They include as special case the absolute value of a volume form.

CHAPTER IX

Stokes' Theorem

Throughout the chapter, manifolds may have a boundary.

IX, §1. STOKES' THEOREM FOR A RECTANGULAR SIMPLEX

If X is a manifold and Y a submanifold, then any differential form on X induces a form on Y. We can view this as a very special case of the inverse image of a form, under the embedding (injection) map

$$\text{id:} \ Y \rightarrow X.$$

In particular, if Y has dimension $n-1$, and if $(x_1, \ldots, x_n)$ is a system of coordinates for X at some point of Y such that the points of Y correspond to those coordinates satisfying $x_j = c$ for some fixed number c, and index j, and if the form on X is given in terms of these coordinates by

$$\omega(x) = f(x_1, \ldots, x_n) \, dx_1 \wedge \cdots \wedge dx_n,$$

then the restriction of ω to Y (or the form induced on Y) has the representation

$$f(x_1, \ldots, c, \ldots, x_n) \, dx_1 \wedge \cdots \wedge \widehat{dx_j} \wedge \cdots \wedge dx_n.$$

We should denote this induced form by ω_Y, although occasionally we omit the subscript Y. We shall use such an induced form especially when Y is the boundary of a manifold X.

200

Let

$$R = [a_1, b_1] \times \cdots \times [a_n, b_n]$$

be a rectangle in n-space, that is a product of n closed intervals. The set theoretic boundary of R consists of the union over all $i = 1, \ldots, n$ of the pieces

$$R_i^0 = [a_1, b_1] \times \cdots \times \{a_i\} \times \cdots \times \{a_n, b_n\},$$
$$R_i^1 = [a_1, b_1] \times \cdots \times \{b_i\} \times \cdots \times [a_n, b_n].$$

If

$$\omega(x_1, \ldots, x_n) = f(x_1, \ldots, x_n)\, dx_1 \wedge \cdots \wedge \widehat{dx_j} \wedge \cdots \wedge dx_n$$

is an $(n-1)$-form, and the roof over anything means that this thing is to be omitted, then we define

$$\int_{R^0} \omega = \int_{a_i}^{b_1} \cdots \int_{a_i}^{\widehat{b_i}} \cdots \int_{a_n}^{b_n} f(x_1, \ldots, a_i, \ldots, x_n)\, dx_1 \cdots \widehat{dx_j} \cdots dx_n,$$

if $i = j$, and 0 otherwise. And similarly for the integral over R_i^1. We define the integral over the oriented boundary to be

$$\int_{\partial^0 R} = \sum_{i=1}^{n} (-1)^i \left[\int_{R_i^0} - \int_{R_i^1} \right].$$

Stokes' Theorem for Rectangles. *Let R be a rectangle in an open set U in n-space. Let ω be an $(n-1)$-form on U. Then*

$$\int_R d\omega = \int_{\partial^0 R} \omega.$$

Proof. In two dimensions, the picture looks like this:

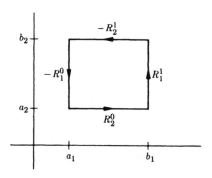

It suffices to prove the assertion when ω is a decomposable form, say

$$\omega(x) = f(x_1, \ldots, x_n)\, dx_1 \wedge \cdots \wedge \widehat{dx_j} \wedge \cdots \wedge dx_n.$$

We then evaluate the integral over the boundary of R. If $i \neq j$, then it is clear that

$$\int_{R_i^0} \omega = 0 = \int_{R_i^1} \omega,$$

so that

$$\int_{\partial^0 R} \omega =$$

$$(-1)^j \int_{a_1}^{b_1} \cdots \widehat{\int_{a_j}^{b_j}} \cdots \int_{a^n}^{b_n} [f(x_1, \ldots, a_j, \ldots, x_n) - f(x_1, \ldots, b_j, \ldots, x_n)]\, dx_1 \cdots \widehat{dx_j} \cdots dx_n.$$

On the other hand, from the definitions we find that

$$d\omega(x) = \left(\frac{\partial f}{\partial x_1}\, dx_1 + \cdots + \frac{\partial f}{\partial x_n}\, dx_n \right) \wedge dx_1 \wedge \cdots \wedge \widehat{dx_j} \wedge \cdots \wedge dx_n$$

$$= (-1)^{j-1} \frac{\partial f}{\partial x_j}\, dx_1 \wedge \cdots \wedge dx_n.$$

(The $(-1)^{j-1}$ comes from interchanging dx_j with $dx_1, \ldots, dx_{j-1}$. All other terms disappear by the alternating rule.)

Integrating $d\omega$ over R, we may use repeated integration and integrate $\partial f/\partial x_j$ with respect to x_j first. Then the fundamental theorem of calculus for one variable yields

$$\int_{a_j}^{b_j} \frac{\partial f}{\partial x_j}\, dx_j = f(x_1, \ldots, b_j, \ldots, x_n) - f(x_1, \ldots, a_j, \ldots, x_n).$$

We then integrate with respect to the other variables, and multiply by $(-1)^{j-1}$. This yields precisely the value found for the integral of ω over the oriented boundary $\partial^0 R$, and proves the theorem.

Remark. Stokes' theorem for a rectangle extends at once to a version in which we parametrize a subset of some space by a rectangle. Indeed, if $\sigma: R \to V$ is a C^1 map of a rectangle of dimension n into an open set V in $\mathbf{R}^N$, and if ω is an $(n-1)$-form in V, we may define

$$\int_\sigma d\omega = \int_R \sigma^* d\omega.$$

One can define

$$\int_{\partial\sigma}\omega = \int_{\partial^0 R}\sigma^*\omega,$$

and then we have a formula

$$\int_\sigma d\omega = \int_{\partial\sigma}\omega,$$

In the next section, we prove a somewhat less formal result.

IX, §2. STOKES' THEOREM ON A MANIFOLD

Theorem 2.1. *Let X be an oriented manifold of class C^2, dimension n, and let ω be an $(n-1)$-form on X, of class C^1. Assume that ω has compact support. Then*

$$\int_X d\omega = \int_{\partial X}\omega.$$

Proof. Let $\{\alpha_i\}_{i\in I}$ be a partition of unity, of class C^2. Then

$$\sum_{i\in I}\alpha_i\omega = \omega,$$

and this sum has only a finite number of non-zero terms since the support of ω is compact. Using the additivity of the operation d, and that of the integral, we find

$$\int_X d\omega = \sum_{i\in I}\int_X d(\alpha_i\omega).$$

Suppose that α_i has compact support in some open set V_i of X and that we can prove

$$\int_{V_i} d(\alpha_i\omega) = \int_{V_i\cap\partial X}\alpha_i\omega,$$

in other words we can prove Stokes' theorem locally in V_i. We can write

$$\int_{V_i\cap\partial X}\alpha_i\omega = \int_{\partial X}\alpha_i\omega,$$

and similarly

$$\int_{V_i} d(\alpha_i \, \omega) = \int_X d(\alpha_i \, \omega).$$

Using the additivity of the integral once more, we get

$$\int_X d\omega = \sum_{i \in I} \int_X d(\alpha_i \, \omega) = \sum_{i \in I} \int_{\partial X} \alpha_i \, \omega = \int_{\partial X} \omega,$$

which yields Stokes' theorem on the whole manifold. Thus our argument with partitions of unity reduces Stokes' theorem to the local case, namely it suffices to prove that for each point of X these exists an open neighborhood V such that if ω has compact support in V, then Stokes' theorem holds with X replaced by V. We now do this.

If the point is not a boundary point, we take an oriented chart (U, φ) at the point, containing an open neighborhood V of the point, satisfying the following conditions: φU is an open ball, and φV is the interior of a rectangle, whose closure is contained in φU. If ω has compact support in V, then its local representation in φU has compact support in φV. Applying Stokes' theorem for rectangles as proved in the preceding section, we find that the two integrals occurring in Stokes' formula are equal to 0 in this case (the integral over an empty boundary being equal to 0 by convention).

Now suppose that we deal with a boundary point. We take an oriented chart (U, φ) at the point, having the following properties. First, φU is described by the following inequalities in terms of local coordinates $(x_1, \dots, x_n)$:

$$-2 < x_1 \leqq 1 \quad \text{and} \quad -2 < x_j < 2 \quad \text{for} \quad j = 2, \dots, n.$$

Next, the given point has coordinates $(1, 0, \dots, 0)$, and that part of U on the boundary of X, namely $U \cap \partial X$, is given in terms of these coordinates by the equation $x_1 = 1$. We then let V consist of those points whose local coordinates satisfy

$$0 < x_1 \leqq 1 \quad \text{and} \quad -1 < x_j < 1 \quad \text{for} \quad j = 2, \dots, n.$$

If ω has compact support in V, then ω is equal to 0 on the boundary of the rectangle R equal to the closure of φV, except on the face given by $x_1 = 1$, which defines that part of the rectangle corresponding to $\partial X \cap V$. Thus the support of ω looks like the shaded portion of the following picture.

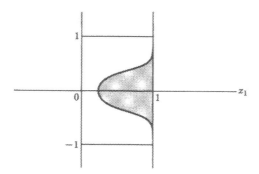

In the sum giving the integral over the boundary of a rectangle as in the previous section, only one term will give a non-zero contribution, corresponding to $i = 1$, which is

$$(-1)\left[\int_{R_1^0} \omega - \int_{R_1^1} \omega\right].$$

Furthermore, the integral over R_1^0 will also be 0, and in the contribution of the integral over R_1^1, the two minus signs will cancel, and yield the integral of ω over the part of the boundary lying in V, because our charts are so chosen that $(x_2, \dots, x_n)$ is an oriented system of coordinates for the boundary. Thus we find

$$\int_V d\omega = \int_{V \cap \partial X} \omega,$$

which proves Stokes' theorem locally in this case, and concludes the proof of Theorem 2.7.

Corollary 2.2. *Suppose X is an oriented manifold without boundary, and ω has compact support. Then*

$$\int_X d\omega = 0.$$

For any number of reasons, some of which we consider in the next section, it is useful to formulate conditions under which Stokes' theorem holds even when the form ω does not have compact support. We shall say that ω has **almost compact support** if there exists a decreasing sequence of open sets $\{U_k\}$ in X such that the intersection

$$\bigcap_{k=1}^{\infty} U_k$$

is empty, and a sequence of C^1 functions $\{g_k\}$, having the following properties:

AC 1. *We have* $0 \leq g_k \leq 1$, $g_k = 1$ *outside* U_k, *and* $g_k\omega$ *has compact support.*

AC 2. *If* μ_k *is the measure associated with* $|dg_k \wedge \omega|$ *on* X, *then*

$$\lim_{k \to \infty} \mu_k(\bar{U}_k) = 0.$$

We then have the following application of Stokes' theorem.

Corollary 2.3. *Let* X *be a* C^2 *oriented manifold, of dimension n, and let* ω *be an* $(n-1)$-*form on* X, *of class* C^1. *Assume that* ω *has almost compact support, and that the measures associated with* $|d\omega|$ *on* X *and* $|\omega|$ *on* ∂X *are finite. Then*

$$\int_X d\omega = \int_{\partial X} \omega.$$

Proof. By our standard form of Stokes' theorem we have

$$\int_{\partial X} g_k\omega = \int_X d(g_k\,\omega) = \int_X dg_k \wedge \omega + \int_X g_k\,d\omega.$$

We estimate the left-hand side by

$$\left| \int_{\partial X} \omega - \int_{\partial X} g_k\,\omega \right| = \left| \int_{\partial X} (1 - g_k)\omega \right| \leq \mu_{|\omega|}(U_k \cap \partial X).$$

Since the intersection of the sets U_k is empty, it follows for a purely measure-theoretic reason that

$$\lim_{k \to \infty} \int_{\partial X} g_k\,\omega = \int_{\partial X} \omega.$$

Similarly,

$$\lim_{k \to \infty} \int_X g_k\,d\omega = \int_X d\omega.$$

The integral of $dg_k \wedge \omega$ over X approaches 0 as $k \to \infty$ by assumption, and the fact that $dg_k \wedge \omega$ is equal to 0 on the complement of $\bar{U}_k$ since g_k is constant on this complement. This proves our corollary.

The above proof shows that the second condition **AC 2** is a very natural one to reduce the integral of an arbitrary form to that of a form with compact support. In the next section, we relate this condition to a question of singularities when the manifold is embedded in some bigger space.

IX, §3. STOKES' THEOREM WITH SINGULARITIES

If X is a compact manifold, then of course every differential form on X has compact support. However, the version of Stokes' theorem which we have given is useful in contexts when we start with an object which is not a manifold, say as a subset of $\mathbf{R}^n$, but is such that when we remove a portion of it, what remains is a manifold. For instance, consider a cone (say the solid cone) as illustrated in the next picture.

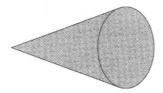

The vertex and the circle surrounding the base disc prevent the cone from being a submanifold of $\mathbf{R}^3$. However, if we delete the vertex and this circle, what remains is a submanifold with boundary embedded in $\mathbf{R}^3$. The boundary consists of the conical shell, and of the base disc (without its surrounding circle). Another example is given by polyhedra, as on the following figure.

The idea is to approximate a given form by a form with compact support, to which we can apply Theorem 2.1, and then take the limit. We shall indicate one possible technique to do this.

The word "boundary" has been used in two senses: The sense of point set topology, and the sense of boundary of a manifold. Up to now, they were used in different contexts so no confusion could arise. We must now make a distinction, and therefore use the word boundary only in its manifold sense. If X is a subset of $\mathbf{R}^N$, we denote its closure by $\bar{X}$ as usual. We call the set-theoretic difference $\bar{X} - X$ the **frontier** of X in $\mathbf{R}^N$, and denote it by $\mathrm{fr}(X)$.

Let X be a submanifold without boundary of $\mathbf{R}^N$, of dimension n. We know that this means that at each point of X there exists a chart for an open neighborhood of this point in $\mathbf{R}^N$ such that the points of X in this chart correspond to a factor in a product. A point P of $\overline{X} - X$ will be called a **regular** frontier point of X if there exists a chart at P in $\mathbf{R}^N$ with local coordinates $(x_1, \ldots, x_N)$ such that P has coordinates $(0, \ldots, 0)$; the points of X are those with coordinates

$$x_{n+1} = \cdots = x_N = 0 \quad \text{and} \quad x_n < 0;$$

and the points of the frontier of X which lie in the chart are those with coordinates satisfying

$$x_n = x_{n+1} = \cdots = x_N = 0.$$

The set of all regular frontier points of X will be denoted by ∂X, and will be called the **boundary** of X. We may say that $X \cup \partial X$ is a submanifold of $\mathbf{R}^N$, possibly with boundary.

A point of the frontier of X which is not regular will be called **singular**. It is clear that the set of singular points is closed in $\mathbf{R}^N$. We now formulate a version of Theorem 2.1 when ω does not necessarily have compact support in $X \cup \partial X$. Let S be a subset of $\mathbf{R}^N$. By a **fundamental sequence** of open neighborhoods of S we shall mean a sequence $\{U_k\}$ of open sets containing S such that, if W is an open set containing S, then $U_k \subset W$ for all sufficiently large k.

Let S be the set of singular frontier points of X and let ω be a form defined on an open neighborhood of $\overline{X}$, and having compact support. The intersection of supp ω with $(X \cup \partial X)$ need not be compact, so that we cannot apply Theorem 2.1 as it stands. The idea is to find a fundamental sequence of neighborhods $\{U_k\}$ of S, and a function g_k which is 0 on a neighborhood of S and 1 outside U_k so that $g_k\omega$ differs from ω only inside U_k. We can then apply Theorem 2.1 to $g_k\omega$ and we hope that taking the limit yields Stokes' theorem for ω itself. However, we have

$$\int_X d(g_k\,\omega) = \int_X dg_k \wedge \omega + \int_X g_k\,d\omega.$$

Thus we have an extra term on the right, which should go to 0 as $k \to \infty$ if we wish to apply this method. In view of this, we make the following definition.

Let S be a closed subset of $\mathbf{R}^N$. We shall say that S is **negligible for** X if there exists an open neighborhood U of S in $\mathbf{R}^N$, a fundamental sequence of open neighborhoods $\{U_k\}$ of S in U, with $\overline{U}_k \subset U$, and a sequence of C^1 functions $\{g_k\}$, having the following properties.

NEG 1. *We have* $0 \leq g_k \leq 1$. *Also,* $g_k(x) = 0$ *for* x *in some open neighborhood of* S, *and* $g_k(x) = 1$ *for* $x \notin U_k$.

NEG 2. *If* ω *is an* $(n-1)$-*form of class* C^1 *on* U, *and* μ_k *is the measure associated with* $|dg_k \wedge \omega|$ *on* $U \cap X$, *then* μ_k *is finite for large* k, *and*

$$\lim_{k \to \infty} \mu_k(U \cap X) = 0.$$

From our first condition, we see that $g_k \omega$ vanishes on an open neighborhood of S. Since $g_k = 1$ on the complement of $\bar{U}_k$, we have $dg_k = 0$ on this complement, and therefore our second condition implies that the measures induced on X near the singular frontier by $|dg_k \wedge \omega|$ (for $k = 1, 2, \ldots$), are concentrated on shrinking neighborhoods and tend to 0 as $k \to \infty$.

Theorem 3.1 (Stokes' Theorem with Singularities). *Let* X *be an oriented,* C^3 *submanifold without boundary of* $\mathbf{R}^N$. *Let* $\dim X = n$. *Let* ω *be an* $(n-1)$-*form of class* C^1 *on an open neighborhood of* $\bar{X}$ *in* $\mathbf{R}^N$, *and with compact support. Assume that*:

(i)   *If* S *is the set of singular points in the frontier of* X, *then* $S \cap \operatorname{supp} \omega$ *is negligible for* X.

(ii) *The measures associated with* $|d\omega|$ *on* X, *and* $|\omega|$ *on* ∂X, *are finite.*

Then

$$\int_X d\omega = \int_{\partial X} \omega.$$

Proof. Let U, $\{U_k\}$, and $\{g_k\}$ satisfy conditions **NEG 1** and **NEG 2**. Then $g_k \omega$ is 0 on an open neighborhood of S, and since ω is assumed to have compact support, one verifies immediately that

$$(\operatorname{supp} g_k \, \omega) \cap (X \cup \partial X)$$

is compact. Thus Theorem 2.1 is applicable, and we get

$$\int_{\partial X} g_k \, \omega = \int_X d(g_k \, \omega) = \int_X dg_k \wedge \omega + \int_X g_k \, d\omega.$$

We have

$$\left| \int_{\partial X} \omega - \int_{\partial X} g_k \, \omega \right| \leq \left| \int_{\partial X} (1 - g_k)\omega \right|$$

$$\leq \int_{U_k \cap \partial X} 1 \, d\mu_{|\omega|} = \mu_{|\omega|}(U_k \cap \partial X).$$

Since the intersection of all sets $U_k \cap \partial X$ is empty, it follows from purely measure-theoretic reasons that the limit of the right-hand side is 0 as $k \to \infty$. Thus

$$\lim_{k \to \infty} \int_{\partial X} g_k \, \omega = \int_{\partial X} \omega.$$

For similar reasons, we have

$$\lim_{k \to \infty} \int_X g_k \, d\omega = \int_X d\omega.$$

Our second assumption **NEG 2** guarantees that the integral of $dg_k \wedge \omega$ over X approaches 0. This proves our theorem.

Criterion 1. *Let* S, T *be compact negligible sets for a submanifold* X *of* $\mathbf{R}^N$ *(assuming* X *without boundary). Then the union* $S \cup T$ *is negligible for* X.

Proof. Let U, $\{U_k\}$, $\{g_k\}$ and V, $\{V_k\}$, $\{h_k\}$ be triples associated with S and T respectively as in condition **NEG 1** and **NEG 2** (with V replacing U and h replacing g when T replaces S). Let

$$W = U \cup V, \qquad W_k = U_k \cup V_k, \qquad \text{and} \qquad f_k = g_k \, h_k.$$

Then the open sets $\{W_k\}$ form a fundamental sequence of open neighborhoods of $S \cup T$ in W, and **NEG 1** is trivially satisfied. As for **NEG 2**, we have

$$d(g_k \, h_k) \wedge \omega = h_k \, dg_k \wedge \omega + g_k \, dh_k \wedge \omega,$$

so that **NEG 2** is also trivially satisfied, thus proving our criterion.

Criterion 2. *Let* X *be an open set, and let* S *be a compact subset in* $\mathbf{R}^n$. *Assume that there exists a closed rectangle* R *of dimension* $m \leqq n - 2$ *and a* C^1 *map* $\sigma \colon R \to \mathbf{R}^n$ *such that* $S = \sigma(R)$. *Then* S *is negligible for* X.

Before giving the proof, we make a couple of simple remarks. First, we could always take $m = n - 2$, since any parametrization by a rectangle of dimension $< n - 2$ can be extended to a parametrization by a rectangle of dimension $n - 2$ simply by projecting away coordinates. Second, by our first criterion, we see that a finite union of sets as described above, that is parametrized smoothly by rectangles of codimension $\geqq 2$, are negligible. Third, our Criterion 2, combined with the first criterion, shows that negligibility in this case is local, that is we can subdivide a rectangle into small pieces.

We now prove Criterion 2. Composing σ with a suitable linear map, we may assume that R is a unit cube. We cut up each side of the cube into k equal segments and thus get k^m small cubes. Since the derivative of σ is bounded on a compact set, the image of each small cube is contained in an n-cube in $\mathbf{R}^N$ of radius $\leq C/k$ (by the mean value theorem), whose n-dimensional volume is $\leq (2C)^n/k^n$. Thus we can cover the image by small cubes such that the sum of their n-dimensional volumes is

$$\leq (2C)^n/k^{n-m} \leq (2C)^n/k^2.$$

Lemma 3.2. *Let S be a compact subset of $\mathbf{R}^n$. Let U_k be the open set of points x such that $d(x, S) < 2/k$. There exists a C^∞ function g_k on $\mathbf{R}^N$ which is equal to 0 in some open neighborhood of S, equal to 1 outside U_k, $0 \leq g_k \leq 1$, and such that all partial derivatives of g_k are bounded by $C_1 k$, where C_1 is a constant depending only on n.*

Proof. Let φ be a C^∞ function such that $0 \leq \varphi \leq 1$, and

$$\varphi(x) = 0 \quad \text{if} \quad 0 \leq \|x\| \leq \tfrac{1}{2},$$
$$\varphi(x) = 1 \quad \text{if} \quad 1 \leq \|x\|.$$

We use $\|\ \|$ for the sup norm in $\mathbf{R}^n$. The graph of φ looks like this:

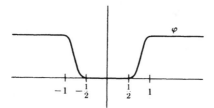

For each positive integer k, let $\varphi_k(x) = \varphi(kx)$. Then each partial derivative $D_i\varphi_k$ satisfies the bound

$$\|D_i\varphi_k\| \leq k\|D_i\varphi\|,$$

which is thus bounded by a constant times k. Let L denote the lattice of integral points in $\mathbf{R}^n$. For each $l \in L$, we consider the function

$$x \mapsto \varphi_k\left(x - \frac{l}{2k}\right).$$

This function has the same shape as φ_k but is translated to the point $l/2k$. Consider the product

$$g_k(x) = \prod \varphi_k\left(x - \frac{l}{2k}\right)$$

taken over all $l \in L$ such that $d(l/2k, S) \leq 1/k$. If x is a point of $\mathbf{R}^n$ such that $d(x, S) < 1/4k$, then we pick an l such that

$$d(x, l/2k) \leq 1/2k.$$

For this l we have $d(l/2, S) < 1/k$, so that this l occurs in the product, and

$$\varphi_k(x - l/2k) = 0.$$

Therefore g_k is equal to 0 in an open neighborhood of S. If, on the other hand, we have $d(x, S) > 2/k$ and if l occurs in the product, that is

$$d(l/2k, S) \leq 1/k,$$

then

$$d(x, l/2k) > 1/k,$$

and hence $g_k(x) = 1$. The partial derivatives of g_k the bounded in the desired manner. This is easily seen, for if x_0 is a point where g_k is not identically 1 in a neighborhood of x_0, then $\|x_0 - l_0/2k\| \leq 1/k$ for some l_0. All other factors $\varphi_k(x - 1/2k)$ will be identically 1 near x_0 unless $\|x_0 - l/2k\| \leq 1/k$. But then $\|l - l_0\| \leq 4$ whence the number of such l is bounded as a function of n (in fact by 9^n). Thus when we take the derivative, we get a sum of a most 9^n terms, each one having a derivative bounded by $C_1 k$ for some constant C_1. This proves our lemma.

We return to the proof of Criterion 2. We observe that when an $(n-1)$-form ω is expressed n terms of its coordinates,

$$\omega(x) = \sum f_j(x)\, dx_1 \wedge \cdots \wedge \widehat{dx_j} \wedge \cdots \wedge dx_n,$$

then the coefficients f_j are bounded on a compact neighborhood of S. We take U_k as in the lemma. Then for k large, each function

$$x \mapsto f_j(x) D_j g_k(x)$$

is bounded on U_k by a bound $C_2 k$, where C_2 depends on a bound for ω, and on the constant of the lemma. The Lebesgue measure of U_k is bounded by C_3/k^2, as we saw previously. Hence the measure of U_k associated with $|dg_k \wedge \omega|$ is bounded by C_4/k, and tends to 0 as $k \to \infty$. This proves our criterion.

As an example, we now state a simpler version of Stokes' theorem, applying our criteria.

Theorem 3.3. *Let X be an open subset of $\mathbf{R}^n$. Let S be the set of singular points in the closure of X, and assume that S is the finite union of C^1 images of m-rectangles with $m \leqq n - 2$. Let ω be an $(n - 1)$-form defined on an open neighborhood of $\overline{X}$. Assume that ω has compact support, and that the measure associated with $|\omega|$ on ∂X and with $|d\omega|$ on X are finite. Then*

$$\int_X d\omega = \int_{\partial X} \omega.$$

Proof. Immediate from our two criteria and Theorem 3.2.

We can apply Theorem 3.3 when, for instance, X is the interior of a polyhedron, whose interior is open in $\mathbf{R}^n$. When we deal with a submanifold X of dimension n, embedded in a higher dimensional space $\mathbf{R}^N$, then one can reduce the analysis of the singular set to Criterion 2 provided that there exists a finite number of charts for X near this singular set on which the given form ω is bounded. This would for instance be the case with the surface of our cone mentioned at the beginning of the section. Criterion 2 is also the natural one when dealing with manifolds defined by algebraic inequalities. By using Hironaka's resolution of singularities, one can parametrize a compact set of algebraic singularities as in Criterion 2.

Finally, we note that the condition that ω have compact support in an open neighborhood of $\overline{X}$ is a very mild condition. If for instance X is a bounded open subset of $\mathbf{R}^n$, then $\overline{X}$ is compact. If ω is any form on some open set containing $\overline{X}$, then we can find another form η which is equal to ω on some open neighborhood of $\overline{X}$ and which has compact support. The integrals of η entering into Stokes' formula will be the same as those of ω. To find η, we simply multiply ω with a suitable C^∞ function which is 1 in a neighborhood of $\overline{X}$ and vanishes a little further away. Thus Theorem 3.3 provides a reasonably useful version of Stokes' theorem which can be applied easily to all the cases likely to arise naturally.

CHAPTER X

Applications of Stokes' Theorem

In this chapter we give a survey of applications of Stokes' theorem, concerning many situations. Some come just from the differential theory, such as the computation of the maximal de Rham cohomology (the space of all forms of maximal degree modulo the subspace of exact forms); some come from Riemannian geometry; and some come from complex manifolds, as in Cauchy's theorem and the Poincaré residue theorem. I hope that the selection of topics will give readers an outlook conducive for further expansion of perspectives. The sections of this chapter are logically independent of each other, so the reader can pick and choose according to taste or need.

X, §1. THE MAXIMAL DE RHAM COHOMOLOGY

Let X be a manifold of dimension n without boundary. Let r be an integer $\geqq 0$. We let $\mathscr{A}^r(X)$ be the **R**-vector space of differential forms on X of degree r. Thus $\mathscr{A}^r(X) = 0$ if $r > n$. If $\omega \in \mathscr{A}^r(X)$, we define the **support** of ω to be the closure of the set of points $x \in X$ such that $\omega(x) \neq 0$.

Examples. If $\omega(x) = f(x) \, dx_1 \wedge \cdots \wedge dx_n$ on some open subset of **R**^{n}, then the support of ω is the closure of the set of x such that $f(x) \neq 0$.

We denote the support of a form ω by $\operatorname{supp}(\omega)$. By definition, the support is closed in X. We are interested in the space of maximal degree forms $\mathscr{A}^n(X)$. Every form $\omega \in \mathscr{A}^n(X)$ is such that $d\omega = 0$. On the other hand, $\mathscr{A}^n(X)$ contains the subspace of **exact** forms, which are defined to

214

be those forms equal to $d\eta$ for some $\eta \in \mathscr{A}^{n-1}(X)$. The factor space is defined to be the **de Rham cohomology** $H^n(X) = H^n(X, \mathbf{R})$. The main theorem of this section can then be formulated.

Theorem 1.1. *Assume that X is compact, orientable, and connected. Then the map*

$$\omega \mapsto \int_X \omega$$

induces an isomorphism of $H^n(X)$ with $\mathbf{R}$ itself. In particular, if ω is in $\mathscr{A}^n(X)$ then there exists $\eta \in \mathscr{A}^{n-1}(X)$ such that $d\eta = \omega$ if and only if

$$\int_X \omega = 0.$$

Actually the hypothesis of compactness on X is not needed. What is needed is compactness on the support of the differential forms. Thus we are led to define $\mathscr{A}^r_c(X)$ to be the vector space of n-forms with compact support. We call a form **compactly exact** if it is equal to $d\eta$ for some $\eta \in \mathscr{A}^{r-1}_c(X)$. We let

$$H^n_c(X) = \text{factor space } \mathscr{A}^n_c(X)/d\mathscr{A}^{n-1}_c(X).$$

Then we have the more general version:

Theorem 1.2. *Let X be a manifold without boundary, of dimension n. Suppose that X is orientable and connected. Then the map*

$$\omega \mapsto \int_X \omega$$

induces an isomorphism of $H^n_c(X)$ with $\mathbf{R}$ itself.

Proof. By Stokes' theorem (Chapter IX, Corollary 2.2) the integral vanishes on exact forms (with compact support), and hence induces an $\mathbf{R}$-linear map of $H^n_c(X)$ into $\mathbf{R}$. The theorem amounts to proving the converse statement: if

$$\int_X \omega = 0,$$

then there exists some $\eta \in \mathscr{A}^{n-1}_c(X)$ such that $\omega = d\eta$. For this, we first have to prove the result locally in $\mathbf{R}^n$, which we now do.

As a matter of notation, we let

$$I^n = (0, 1)^n$$

be the open n-cube in $\mathbf{R}^n$. What we want is:

Lemma 1.3. *Let ω be an n-form on I^n, with compact support, and such that*

$$\int_{I^n} \omega = 0.$$

Then there exists a form $\eta \in \mathscr{A}_c^{n-1}(I^{n-1})$ with compact support, such that

$$\omega = d\eta.$$

We will prove Lemma 1.3 by induction, but it is necessary to load to induction to carry it out. So we need to prove a stronger version of Lemma 1.3 as follows.

Lemma 1.4. *Let ω be an $(n-1)$-form on I^{n-1} whose coefficient is a function of n variables $(x_1, \ldots, x_n)$ so*

$$\omega(x) = f(x_1, \ldots, x_n)\, dx_1 \wedge \cdots \wedge dx_{n-1}.$$

(Of course, all functions, like forms, are assumed C^∞.) Suppose that ω has compact support in I^{n-1}. Assume that

$$\int_{I^{n-1}} \omega = 0.$$

Then there exists an $(n-1)$-form η, whose coefficients are C^∞ functions of $x_1, \ldots, x_n$ with compact support such that

$$\omega(x_1, \ldots, x_{n-1}; x_n) = d_{n-1}\, \eta(x_1, \ldots, x_{n-1}; x_n).$$

The symbol d_{n-1} here means the usual exterior derivative taken with respect to the first $n-1$ variables.

Proof. By induction. We first prove the theorem when $n - 1 = 1$. First we carry out the proof leaving out the extra variable, just to see what's going on. So let
$$\omega(x) = f(x)\, dx,$$

where f has compact support in the open interval $(0, 1)$. This means there exists $\epsilon > 0$ such that $f(x) = 0$ if $0 < x \leq \epsilon$ and if $1 - \epsilon \leq x \leq 1$. We assume

$$\int_0^1 f(x)\, dx = 0.$$

Let

$$g(x) = \int_0^x f(t)\,dt.$$

Then $g(x) = 0$ if $0 < x \leq \epsilon$, and also if $1 - \epsilon \leq x \leq 1$, because for instance if $1 - \epsilon \leq x \leq 1$, then

$$g(x) = \int_0^1 f(t)\,dt = 0.$$

Then $f(x)\,dx = dg(x)$, and the lemma is proved in this case. Note that we could have carried out the proof with the extra variable x_2, starting from

$$\omega(x) = f(x_1, x_2)\,dx_1,$$

so that

$$g(x_1, x_2) = \int_0^1 f(t, x_2)\,dt.$$

We can differentiate under the integral sign to verify that g is C^∞ in the pair of variables (x_1, x_2).

Now let $n \geq 3$ and assume that theorem proved for $n - 1$ by induction. To simplify the notation, let us omit the extra variable x_{n+1}, and write

$$\omega(x) = f(x_1, \ldots, x_n)\,dx_1 \wedge \cdots \wedge dx_n,$$

with compact support in I^n. Then there exists $\epsilon > 0$ such that the support of f is contained in the closed cube

$$\bar{I}^n(\epsilon) = [\epsilon, 1 - \epsilon]^n.$$

The following figure illustrates this support in dimension 2.

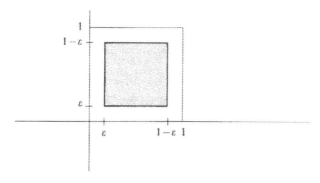

Let ψ be an $(n - 1)$-form on I^{n-1}, $\psi(x) = \psi(x_1, \ldots, x_{n-1})$ such that

$$\int_{I^{n-1}} \psi = 1,$$

and ψ has compact support. Let

$$g(x_n) = \int_{I^{n-1}} f(x_1,\ldots,x_{n-1};x_n)\, dx_1 \wedge \cdots \wedge dx_{n-1}$$

$$= \int_{\bar{I}^{n-1}(\epsilon)} f(x_1,\ldots,x_{n-1};x_n)\, dx_1 \wedge \cdots \wedge dx_{n-1}.$$

Note here that we do have the parameter x_n coming in at the inductive step. Let

$$\mu(x) = f(x)\, dx_1 \wedge \cdots \wedge dx_{n-1} - g(x_n)\psi(x_1,\ldots,x_{n-1}),$$

so

(*) $\mu(x) \wedge dx_n = \omega(x) - g(x_n)\psi(x) \wedge dx_n.$

Then

$$\int_{I^{n-1}} \mu = g(x_n) - g(x_n) = 0.$$

Furthermore, since f has compact support, so does g (look at the figure). By induction, there exists an $(n-1)$-form η, of the first $n-1$ variables, but depending on the parameter x_n, that is

$$\eta(x) = \eta(x_1,\ldots,x_{n-1};x_n)$$

such that

$$\mu(x_1,\ldots,x_{n-1};x_n) = d_{n-1}\eta(x_1,\ldots,x_{n-1};x_n).$$

Here d_{n-1} denotes the exterior derivative with respect to the first $n-1$ variables. Then trivially,

$$\mu(x_1,\ldots,x_{n-1};x_n) \wedge dx_n = d_{n-1}\eta(x_1,\ldots,x_{n-1};x_n) \wedge dx_n$$
$$= d\eta(x),$$

where $d\eta$ is now the exterior derivative taken with respect to all n variables. Hence finally from equation (*) we obtain

(**) $\omega(x) = d\eta(x) + g(x_n)\psi(x_1,\ldots,x_{n-1}) \wedge dx_n.$

To conclude the proof of Lemma 1.3, it suffices to show that the second term on the right of (**) is exact. We are back to a one-variable problem. Let

$$h(x_n) = \int_0^{x_n} g(t)\, dt.$$

Then $dh(x_n) = g(x_n)dx_n$, and h has compact support in the interval $(0, 1)$, just as in the start of the induction. Then

$$d\big(h(x_n)\psi(x_1, \ldots, x_{n-1})\big) = dh(x_n) \wedge \psi(x_1, \ldots, x_{n-1})$$
$$= (-1)^{n-1} g(x_n)\psi(x_1, \ldots, x_{n-1}) \wedge dx_n$$

because $d\psi = 0$. Of course we could have carried along the extra parameter all the way through. This concludes the proof of Lemma 1.3.

We formulate an immediate consequence of Lemma 1.3 directly on the manifold.

Lemma 1.5. *Let U be an open subset of X, isomorphic to I^n. Let $\psi \in \mathscr{A}_c^n(U)$ be such that*

$$\int_U \psi \neq 0.$$

Let $\omega \in \mathscr{A}_c^n(U)$. Then there exists $c \in \mathbf{R}$ and $\eta \in \mathscr{A}_c^{n-1}(U)$ such that

$$\omega - c\psi = d\eta.$$

Proof. We take $c = \int_U \omega \Big/ \int_U \psi$ and apply Lemma 1.3 to $\omega - c\psi$.

Observe that the hypothesis of connectedness has not yet entered the picture. The preceding lemmas were purely local. We now globalize.

Lemma 1.6. *Assume that X is connected and oriented. Let U, ψ be as in Lemma 1.5. Let V be the set of points $x \in X$ having the following property. There exists a neighborhood $U(x)$ of x isomorphic to I^n such that for every $\omega \in \mathscr{A}_c^n(U(x))$ there exist $c \in \mathbf{R}$ and $\eta \in \mathscr{A}_c^{n-1}(X)$ such that*

$$\omega - c\psi = d\eta.$$

Then $V = X$.

Proof. Lemma 1.5 asserts that $V \supset U$. Since X is connected, it suffices to prove that V is both open and closed. It is immediate from the definition of V that V is open, so there remains to prove its closure. Let z be in the closure of V. Let W be a neighborhood of z isomorphic to I^n.

There exists a point $x \in V \cap W$. There exists a neighborhood $U(x)$ as in the definition of V such that $U(x) \subset W$. For instance, we may take

$$U(x) \approx (a_1, b_1) \times \cdots \times (a_n, b_n) \approx I^n$$

with a_i sufficiently close to 0 and b_i sufficiently close to 1, and of course $0 < a_i < b_i$ for $i = 1, \ldots, n$. Let $\psi_1 \in \mathscr{A}_c^n(U(x))$ be such that

$$\int_{U(x)} \psi_1 = 1.$$

Let $\omega \in \mathscr{A}_c^n(W)$. By the definition of V, there exist $c_1 \in \mathbf{R}$ and $\eta_1 \in \mathscr{A}_c^n(X)$ such that

$$\psi_1 - c_1\psi = d\eta_1.$$

By Lemma 1.5, there exists $c_2 \in \mathbf{R}$ and $\eta_2 \in \mathscr{A}_c^n(X)$ such that

$$\omega - c_2\psi_1 = d\eta_2.$$

Then

$$\omega - c_2 c_1 \psi = d(\eta_2 + c_2\eta_1),$$

thus concluding the proof of Lemma 1.6.

We have now reached the final step in the proof of Theorem 1.2, namely we first fix a form $\psi \in \mathscr{A}_c^n(U)$ with $U \approx I^n$ and $\int_X \psi \neq 0$. Let $\omega \in \mathscr{A}_c^n(X)$. It suffices to prove that there exist $c \in \mathbf{R}$ and $\eta \in \mathscr{A}_c^{n-1}(X)$ such that

$$\omega - c\psi = d\eta.$$

Let K be the compact support of ω. Cover K by a finite number of open neighborhoods $U(x_1), \ldots, U(x_m)$ satisfying the property of Lemma 1.6. Let $\{\varphi_i\}$ be a partition of unity subordinated to this covering, so that we can write

$$\omega = \sum \varphi_i\omega.$$

Then each form $\varphi_i\omega$ has support in some $U(x_j)$. Hence by Lemma 1.6, there exist $c_i \in \mathbf{R}$ and $\eta_i \in \mathscr{A}_c^{n-1}(X)$ such that

$$\varphi_i\omega - c_i\psi = d\eta_i,$$

whence $\omega - c\psi = d\eta$, with $c = \sum c_i$ and $\eta = \sum \eta_i$. This concludes the proof of Theorems 1.1 and 1.2.

X, §2. VOLUME FORMS AND THE DIVERGENCE

Let V be a euclidean vector space over $\mathbf{R}$, of dimension n. We assume given a positive definite symmetric scalar product g, denoted by

$$(v, w) \mapsto \langle v, w \rangle_g = g(v, w) \qquad \text{for } v, w \in V.$$

The space $\bigwedge^n V$ has dimension 1. If $\{e_1, \ldots, e_n\}$ and $\{u_1, \ldots, u_n\}$ are orthonormal bases of V, then

$$e_1 \wedge \cdots \wedge e_n = \pm u_1 \wedge \cdots \wedge u_n.$$

Two such orthonormal bases are said to have the **same orientation**, or to be **orientation equivalent**, if the plus sign occurs in the above relation. A choice of an equivalence class of orthonormal bases having the same orientation is defined to be an **orientation** of V. Thus an orientation determines a basis for the one-dimensional space $\bigwedge^n V$ over $\mathbf{R}$. Such a basis will be called a **volume**. There exists a unique n-form Ω on V (alternating), also denoted by vol_g, such that for every oriented orthonormal basis $\{e_1, \ldots, e_n\}$ we have

$$\Omega(e_1, \ldots, e_n) = 1.$$

Conversely, given a non-zero n-form Ω on V, all orthonormal bases $\{e_1, \ldots, e_n\}$ such that $\Omega(e_1, \ldots, e_n) > 0$ are orientation equivalent, and on such bases Ω has a constant value.

Let (X, g) be a Riemannian manifold. By an **orientation** of (X, g) we mean a choice of a volume form Ω, and an orientation of each tangent space $T_x X$ ($x \in X$) such that for any oriented orthonormal basis $\{e_1, \ldots, e_n\}$ of $T_x X$ we have

$$\Omega_x(e_1, \ldots, e_n) = 1.$$

The form gives a coherent way of making the orientations at different points compatible. It is an exercise to show that if (X, g) has such an orientation, and X is connected, then (X, g) has exactly two orientations. By an **oriented chart**, with coordinates $x_1, \ldots, x_n$ in $\mathbf{R}^n$, we mean a chart such that with respect to these coordinates, the form has the representation

$$\Omega(x) = \varphi(x)\, dx_1 \wedge \cdots \wedge dx_n$$

with a function φ which is positive at every point of the chart. We call Ω the **Riemannian volume form**, and also denote it by vol_g, so

$$\mathrm{vol}_g(x) = \Omega(x) = \Omega_x.$$

We return to our vector space V, with positive definite metric g, and oriented.

Proposition 2.1. *Let* $\Omega = \mathrm{vol}_g$. *Then for all* n-*tuples of vectors* $\{v_1, \ldots, v_n\}$ *and* $\{w_1, \ldots, w_n\}$ *in* V, *we have*

$$\Omega(v_1, \ldots, v_n)\Omega(w_1, \ldots, w_n) = \det\langle v_i, w_j\rangle_g.$$

In particular,

$$\Omega(v_1, \ldots, v_n)^2 = \det\langle v_i, v_j\rangle_g.$$

Proof. The determinant on the right side of the first formula is multilinear and alternating in each n-tuple $\{v_1, \ldots, v_n\}$ and $\{w_1, \ldots, w_n\}$. Hence there exists a number $c \in \mathbf{R}$ such that

$$\det\langle v_i, w_j\rangle_g = c\,\Omega(v_1, \ldots, v_n)\Omega(w_1, \ldots, w_n)$$

for all such n-tuples. Evaluating on an oriented orthonormal basis shows that $c = 1$, thus proving the proposition.

Applying Proposition 2.1 to an oriented Riemannian manifold yields:

Proposition 2.2. *Let* (X, g) *be an oriented Riemannian manifold. Let* $\Omega = \mathrm{vol}_g$. *For all vector fields* $\{\xi_1, \ldots, \xi_n\}$ *and* $\{\eta_1, \ldots, \eta_n\}$ *on* X, *we have*

$$\Omega(\xi_1, \ldots, \xi_n)\Omega(\eta_1, \ldots, \eta_n) = \det\langle \xi_i, \eta_j\rangle_g.$$

In particular,

$$\Omega(\xi_1, \ldots, \xi_n)^2 = \det\langle \xi_i, \xi_j\rangle_g.$$

Furthermore, if $\xi^\vee$ *denotes the one-form dual to* ξ (*characterized by* $\xi^\vee(\eta) = \langle \xi, \eta\rangle_g$ *for all vector fields* η), *then*

$$\Omega(\xi_1, \ldots, \xi_n)\Omega = \xi_1^\vee \wedge \cdots \wedge \xi_n^\vee.$$

This last formula is merely an application of the definition of the wedge product of forms, taking into account the preceding formulas concerning the determinant.

At a point, the space of n-forms is 1-dimensional. Hence any n-form on a Riemannian manifold can be written as a product $\varphi\Omega$ where φ is a function and Ω is the Riemannian volume form.

If ξ is a vector field, then $\Omega \circ \xi$ is an $(n-1)$-form, and so there exists a function φ such that

$$d(\Omega \circ \xi) = \varphi\Omega.$$

We call φ the **divergence** of ξ with respect to Ω, or with respect to the Riemannian metric. We denote it by $\operatorname{div}_\Omega \xi$ or simply $\operatorname{div} \xi$. Thus by definition,

$$d(\Omega \circ \xi) = (\operatorname{div} \xi)\Omega.$$

Example. Looking back at Chapter V, §3 we see that if

$$\Omega(x) = dx_1 \wedge \cdots \wedge dx_n$$

is the canonical form on $\mathbf{R}^n$ and ξ is a vector field, $\xi = \sum \varphi_i u_i$ where $\{u_1, \ldots, u_n\}$ are the standard unit vectors, and φ_i are the coordinate functions, then

$$\operatorname{div}_\Omega \xi = \sum_{i=1}^n \frac{\partial \varphi_i}{\partial x_i}.$$

For the formula with a general metric, see Proposition 2.5.

On 1-forms, we define the operator

$$d^*: \mathscr{A}^1(X) \to \mathscr{A}^0(X)$$

by duality, that is if $\lambda^\vee$ denotes the vector field corresponding to λ under the Riemannian metric, then we define

$$d^*\lambda = -\operatorname{div} \lambda^\vee.$$

Let us define the **Laplacian** or **Laplace operator** *on functions* by the formula

$$\Delta = d^*d = -\operatorname{div} \circ \operatorname{grad}.$$

Proposition 2.3. *For functions* φ, ψ *we have*

$$\Delta(\varphi\psi) = \varphi\Delta\psi + \psi\Delta\varphi - 2\langle d\varphi, d\psi \rangle_g.$$

Proof. The routine gives:

$$\begin{aligned}
\Delta(\varphi\psi) = d^*d(\varphi\psi) &= d^*(\psi \, d\varphi + \varphi \, d\psi) \\
&= -\operatorname{div}(\psi\xi_{d\varphi}) - \operatorname{div}(\varphi\xi_{d\psi}) \\
&= -\psi \operatorname{div} \xi_{d\varphi} - (d\psi)\xi_{d\varphi} - \varphi \operatorname{div} \xi_{d\psi} - (d\varphi)\xi_{d\psi} \\
&= \psi\Delta\varphi + \varphi\Delta\psi - 2\langle d\varphi, d\psi \rangle_g
\end{aligned}$$

as was to be shown.

Recall that

$$\langle d\varphi, d\psi \rangle_g = \langle \operatorname{grad} \varphi, \operatorname{grad} \psi \rangle_g,$$

so there is an alternative expression for the last term in the formula.

We shall tabulate some formulas concerning the gradient. For simplicity of notation, we shall omit the subscript g in the scalar product, because we now fix g. We shall also write simply gr φ instead of $\operatorname{grad}_g \varphi$.

gr 1. For functions φ, ψ we have

$$\operatorname{gr}(\varphi\psi) = \varphi \operatorname{gr}(\psi) + \psi \operatorname{gr}(\varphi).$$

gr 2. The map $\varphi \mapsto (\operatorname{gr}(\varphi))/\varphi = \varphi^{-1} \operatorname{gr}(\varphi)$ is a homomorphism, from the multiplicative group of functions never 0, to the additive group of functions. In particular, for a positive function φ,

$$2\varphi^{-1/2} \operatorname{gr}(\varphi^{1/2}) = \varphi^{-1} \operatorname{gr}(\varphi) = \operatorname{gr} \log \varphi$$

because $d \log \varphi = \varphi^{-1} d\varphi$.

gr 3. $$\operatorname{gr}(\varphi) \cdot \psi = \operatorname{gr}(\psi) \cdot \varphi = \langle \operatorname{gr}(\varphi), \operatorname{gr}(\psi) \rangle_g.$$

We use these formulas to give two versions of certain operators which arise in practice. For any function φ, we write for the Lie derivative

$$[\operatorname{gr} \varphi] = \mathscr{L}_{\operatorname{gr} \varphi}.$$

Corollary 2.4. *Let δ be a positive function. Then*

$$\Delta - [\operatorname{gr} \log \delta] = \delta^{-1/2}\Delta \circ \delta^{1/2} - \delta^{-1/2}\Delta(\delta^{1/2}).$$

Proof. For a function ψ, by Proposition 2.3,

$$(\Delta \circ \delta^{1/2})\psi = \Delta(\delta^{1/2}\psi)$$
$$= \delta^{1/2}\Delta\psi + \psi\Delta(\delta^{1/2}) - 2(\operatorname{gr} \delta^{1/2}) \cdot \psi.$$

We apply the right side of the equality to be proved to a function ψ. We use the formula just derived, mutliplied by $\delta^{-1/2}$. The term $\delta^{-1/2}\Delta(\delta^{1/2})\psi$ cancels, and we obtain

$$(\text{right side})(\psi) = \Delta\psi - 2\delta^{-1/2}(\operatorname{gr} \delta^{1/2}) \cdot \psi.$$

We use **gr 2** to conclude the proof.

General definition of the divergence

Although the most important case of the divergence is on a Riemannian manifold, some properties are most clearly expressed in a more general case which we now describe. Let T be a vector space of finite dimension n over $\mathbf{R}$. Then $\bigwedge^n T$ is of dimension 1, and will be called the **determinant** of T, so by definition,

$$\det T = \bigwedge\nolimits^{\max} T = \bigwedge\nolimits^n T.$$

Observe that we also have $\det T^\vee$. A non-zero element of $\det T^\vee$ will be called a **volume form** on T.

The vector space of sections of $\bigwedge^n T^\vee X$ on a manifold X of dimension n is also a module over the ring of functions. By a **volume form on X** we mean section which is nowhere 0, so a volume form is a basis for this space over the ring of functions. Instead of saying that Ω is a volume form, one may also say that Ω is **non-singular**. If Ψ is any n-form on X, then there exists a function f such that $\Psi = f\Omega$. So let Ω be a volume form. Let ξ be a vector field on X. We define the **divergence of ξ with respect to Ω** just as we did for the Riemannian volume form, namely $\operatorname{div}_\Omega(\xi)$ is defined by the property

DIV 1. $d(\Omega \circ \xi) = \big(\operatorname{div}_\Omega(\xi)\big)\Omega.$

From Chapter V, Proposition 5.3, **LIE 1**, we also have the equivalent defining property

DIV 2. $\mathscr{L}_\xi \Omega = \big(\operatorname{div}_\Omega(\xi)\big)\Omega.$

Directly from **DIV 2** and **LIE 2**, we get for any functions φ, f:

DIV 3. $\operatorname{div}_\Omega(\varphi\xi) = \varphi \operatorname{div}_\Omega(\xi) + \xi \cdot \varphi.$

DIV 4. $df \wedge (\Omega \circ \xi) = (\xi \cdot f)\Omega.$

Proof. First we have $\mathscr{L}_\xi(f\Omega) = (\xi \cdot f)\Omega + f \operatorname{div}_\Omega(\xi)\Omega$, and second,

$$\mathscr{L}_\xi(f\Omega) = d(f\Omega \circ \xi) = df \wedge (\Omega \circ \xi) + f d(\Omega \circ \xi)$$
$$= df \wedge (\Omega \circ \xi) + f \operatorname{div}_\Omega(\xi)\Omega.$$

Then **DIV 4** follows from these two expressions.

One can define an orientation on the general vector space T depending on the non-singular form Ω. Of course in general, we don't have the notion of orthogonality. But we say that a basis $\{v_1, \ldots, v_n\}$ of

T is **positively oriented**, or simply **oriented**, with respect to Ω if $\Omega(v_1, \ldots, v_n) > 0$. Let Ω, Ψ be volume forms. We say that they have the **same orientation**, or that they are **positive with respect to each other**, if there exists a positive function h such that $\Omega = h\Psi$. Forms with the same orientation define the same orientation on bases. A manifold which admits a volume form is said to be **orientable**, and the class of volume forms having the same orientation is said to define the **orientation**.

Let δ be a positive function on X, and let Ψ be a volume form. Then:

DIV 5. $\text{div}_{\delta\Psi}(\xi) = (\xi \cdot \log \delta) + \text{div}_\Psi(\xi)$.

Proof. By Proposition 5.3 of Chapter V, **LIE 1**, we have

$$d(\delta\Psi \circ \xi) = \mathscr{L}_\xi(\delta\Psi) = (\xi \cdot \delta)\delta^{-1}\delta\Psi + \delta\mathscr{L}_\xi(\Psi)$$
$$= (\xi \cdot \log \delta)(\delta\Psi) + \delta \, \text{div}_\Psi(\xi)\Psi,$$

which proves the formula.

The divergence in a chart

Next we obtain an expression for the divergence in a chart. Let U be an open set of a chart for X in $\mathbf{R}^n$ with its standard unit vectors $u_1, \ldots, u_n$. There exists a function δ never 0 on U such that in this chart,

$$\Omega = \delta \, dx_1 \wedge \cdots \wedge dx_n.$$

Suppose U is connected. Then we have $\delta > 0$ on U or $\delta < 0$ on U since Ω is assumed non-singular. For simplicity, assume $\delta > 0$.

Example. If $\Omega = \Omega_g$ is the Riemannian volume form, then

$$\delta = (\det g)^{1/2}.$$

In other words,

$$\Omega_g(x) = \left(\det g(x)\right)^{1/2} dx_1 \wedge \cdots \wedge dx_n.$$

Here $g(x)$ denotes the matrix representing g with respect to the standard basis of $\mathbf{R}^n$.

We write ξ in the chart U as a linear combination

$$\xi = \sum \varphi_i u_i$$

with coordinate functions $\varphi_1, \ldots, \varphi_n$. We let ∂_i be the i-th partial derivative. We write the coordinate vector of ξ vertically, that is

$$\Phi = \Phi_\xi = \begin{pmatrix} \varphi_1 \\ \vdots \\ \varphi_n \end{pmatrix}.$$

We let ${}^t\mathbf{D}_\Omega$ be the *row* vector of operators

$${}^t\mathbf{D}_\Omega = (\partial_1 + \partial_1 \log \delta, \ldots, \partial_n + \partial_n \log \delta).$$

Proposition 2.5.

$$\operatorname{div}_\Omega \xi = \delta^{-1} \sum \partial_i(\delta \varphi_i)$$

$$= \sum \partial_i \varphi_i + \sum (\partial_i \log \delta) \varphi_i.$$

In matrix form,

$$\operatorname{div}_\Omega \xi = {}^t\mathbf{D}_\Omega \Phi_\xi \quad \text{or also} \quad \operatorname{div}_\Omega = \delta^{-1} \, {}^t D \circ \delta.$$

Proof. We have

$$(\Omega \circ \xi)(u_1, \ldots, \hat{u}_i, \ldots, u_n) = \Omega(\xi, u_1, \ldots, \hat{u}_i, \ldots, u_n)$$

$$= (-1)^{i-1} \Omega(u_1, \ldots, \xi, \ldots, u_n)$$

$$= (-1)^{i-1} \delta \varphi_i.$$

Hence

$$(\Omega \circ \xi) = \sum (-1)^{i-1} \delta_{\varphi_i} \, dx_1 \wedge \cdots \wedge \widehat{dx_i} \wedge \cdots \wedge dx_n,$$

and since $ddx_j = 0$ for all j, we obtain

$$d(\Omega \circ \xi) = \sum (-1)^{i-1} \partial_i(\delta \varphi_i) \, dx_i \wedge dx_1 \wedge \cdots \wedge \widehat{dx_i} \wedge \cdots \wedge dx_n$$

$$= \sum \partial_i(\delta \varphi_i) \, dx_1 \wedge \cdots \wedge dx_n$$

$$= \delta^{-1} \sum \partial_i(\delta \varphi_i) \Omega.$$

This proves the proposition.

We return to the gradient, for which we give an expression in local coordinates, with an application to the Laplacian.

Proposition 2.6. *Let* $\mathrm{gr}(\psi) = \sum \varphi_i u_i$. *Let* $g(x)$ *be the* $n \times n$ *matrix representing the metric at a point* x. *Then the coordinate vector of* $\mathrm{gr}(\psi)$ *is*

$$\Phi = \begin{pmatrix} \varphi_1 \\ \vdots \\ \varphi_n \end{pmatrix} = g(x)^{-1} \begin{pmatrix} \partial_1 \psi \\ \vdots \\ \partial_n \psi \end{pmatrix}.$$

In other words,

$$\Phi = g^{-1} \partial \psi,$$

where ∂ *is the vector differential operator such that* ${}^t\partial = (\partial_1, \ldots, \partial_n)$.

Proof. By definition,

$$\langle \mathrm{gr}(\psi), u_j \rangle_g = (d\psi)(u_j) = \partial_j \psi.$$

The left side is equal to $\langle \mathrm{gr}(\psi), g(x)u_j \rangle$ at a point x. Note that here the scalar product is the usual dot product on $\mathbf{R}^n$, without the subscript g. The formula of the proposition then follows at once.

Proposition 2.7. *Let* f *and* ψ *be function, and let* $\mathrm{gr}(\psi) = \sum \varphi_j u_j$ *as in Proposition 2.6. Then*

$$\mathrm{gr}(\psi) \cdot f = \sum_{j=1}^{n} (\partial_j f)\varphi_j.$$

Proof. Since $u_j \cdot f = \partial_j f$, the formula is clear.

From Propositions 2.5 and 2.6, we obtain the coordinate representation of the Laplacian via a matrix:

Proposition 2.8. *On an open set of* $\mathbf{R}^n$, *with metric matrix* g, $\delta = (\det g)^{1/2}$, *and Laplacian* Δ_g, *we have*

$$- \Delta_g = \mathrm{div}_g \, \mathrm{gr}_g = {}^t \mathbf{D}_g g^{-1} \partial$$
$$= \delta^{-1} \, {}^t \partial \delta_g^{-1} \partial.$$

Here, $\mathbf{D}_g$ *abbreviates* $\mathbf{D}_{\Omega_g}$, *and* div_g *abbreviates* div_{Ω_g}.

Putting all the indices in, we get

(1) $$-\Delta_g f = \delta^{-1} \sum_i \partial_i \left(\delta \sum_j g^{ij} \partial_j f \right)$$

where in classical notation, $g^{-1}(x)$ is the matrix $\left(g^{ij}(x)\right)$ for $x \in \mathbf{R}^n$. Using the rule for the derivative of a product, we write (1) in the form

$$(2) \qquad -\Delta_g f = \sum_{i,j=1}^{n} g^{ij}\, \partial_i \partial_j f + L_g f,$$

where L_g is a first-order differential operator, that is a linear combination of the partials $\partial_1, \ldots, \partial_n$ with coefficients which are functions, depending on g. From this expression, we see that the matrix $g^{-1} = (g^{ij})$ is the matrix of the second-order term, quadratic in the partials ∂_i, ∂_j. Hence we obtain:

Theorem 2.9. *Let X be a Riemannian manifold. Then the Laplacian determines the metric, i.e. if two Riemannian metrics have the same Laplacian, they are equal. If $F: X \to Y$ is a differential isomorphism of Riemannian manifolds, and F maps Δ_X on Δ_Y, that is F commutes with the Laplacians, then F is an isometry.*

Note that the second statement about the differential isomorphism is just a piece of functorial abstract nonsense, in light of the first statement. Indeed, F maps the metric g_X to a metric F_*g_X on Y, and similarly for the Laplacian. By assumption, $F_*\Delta_X = \Delta_Y$. Hence Δ_Y is the Laplacian of g_Y and of F_*g_X, so $g_Y = F_*g_X$ by the first statement in the theorem.

Example. Let $A = \mathbf{R}^+ \times \cdots \times \mathbf{R}^+$ be the product of positive multiplicative groups, taken n times, so we view A as an open subset of $\mathbf{R}^n$. We let a denote the variable in A, so $a = {}^t(a_1, \ldots, a_n)$ with $a_i > 0$. We identify the tangent space $T_a A = T_a$ with $\mathbf{R}^n$, so a vector $v \in T_a$ is an ordinary n-tuple,

$$v = {}^t(c_1, \ldots, c_n) \qquad \text{with} \qquad c_i \in \mathbf{R}.$$

Let g be the metric on A defined by the formula

$$\langle v, v \rangle_a = \sum_{i=1}^{n} c_i^2 / a_i^2.$$

Then g is represented by the diagonal matrix $g(a) = \mathrm{diag}(a_1^{-2}, \ldots, a_n^{-2})$, that is

$$\langle v, v \rangle_a = \langle v, g(a)v \rangle,$$

where the scalar product without indices denotes the standard scalar product on $\mathbf{R}^n$. Then

$$\delta(a) = \det g(a)^{1/2} = \prod_{i=1}^{n} a_i^{-1} = \mathbf{d}(a)^{-1}$$

where $\mathbf{d}(a) = a_1 \cdots a_n$ is the product of the coordinates. Thus for a function ψ on A, we have the explicit determination for the gradient,

$$(1) \qquad (\mathrm{gr}_A \psi)(a) = g(a)^{-1} \partial \psi = {}^t(a_1^2 \partial_1 \psi, \ldots, a_n^2 \partial_n \psi)(a).$$

The Laplacian Δ_A from Proposition 2.8 is seen to be

$$(2) \qquad -\Delta_A = \sum_{i=1}^n a_i \partial_i + \sum_{i=1}^n a_i^2 \partial_i^2.$$

This comes from matrix multiplication,

$$\mathbf{d}(a)(\partial_1, \ldots, \partial_n) \begin{pmatrix} \mathbf{d}(a)^{-1} a_1^2 \partial_1 \\ \vdots \\ \mathbf{d}(a)^{-1} a_n^2 \partial_n \end{pmatrix}.$$

X, §3. THE DIVERGENCE THEOREM

Let X be an oriented manifold of dimension n possibly with boundary, and let Ω be an n-form on X. Let ξ be a vector field on X. Then $d\Omega = 0$, and hence the basic formula for the Lie derivative (Chapter V, Proposition 5.3) shows that

$$\mathscr{L}_\xi \Omega = d(\Omega \circ \xi).$$

Consequently in this case, Stokes' theorem yields:

Theorem 3.1 (Divergence Theorem).

$$\boxed{\int_X \mathscr{L}_\xi \Omega = \int_{\partial X} \Omega \circ \xi.}$$

Remark. Even if the manifold is not orientable, it is possible to use the notion of density to formulate a Stokes theorem for densities. Cf. Loomis–Sternberg [LoS 68] for the formulation, due to Rasala. However, this formulation reduces at once to a local question (using partitions of unity on densities). Since locally every manifold is orientable, and a density then amounts to a differential form, this more general formulation again reduces to the standard one on an orientable manifold.

Suppose that (X, g) is a Riemannian manifold, assumed oriented for simplicity. We let Ω or vol_g be the volume form defined in §2. Let ω be the canonical Riemannian volume form on ∂X for the metric induced by g on the boundary. Let $\mathbf{n}_x$ be a unit vector in the tangent space $T_x(X)$ such that u is perpendicular to $T_x(\partial X)$. Such a unit vector is determined up to sign. Denote by $\mathbf{n}_x^{\vee}$ its dual functional, i.e. the component on the projection along $\mathbf{n}_x$. *We select $\mathbf{n}_x$ with the sign such that*

$$\mathbf{n}_x^{\vee} \wedge \omega(x) = \Omega(x).$$

We then shall call $\mathbf{n}_x$ the **unit outward normal vector** to the boundary at x. In an oriented chart, it looks like this.

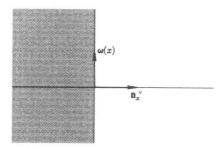

Then by formula **CON 3** of Chapter V, §5 we find

$$\Omega \circ \xi = \langle \mathbf{n}, \xi \rangle \omega - \mathbf{n}^{\vee} \wedge (\omega \circ \xi),$$

and the restriction of this form to ∂X is simply $\langle \mathbf{n}, \xi \rangle \omega$. Thus we get:

Theorem 3.2 (Gauss Theorem). *Let X be a Riemannian manifold. Let ω be the canonical Riemannian volume form on ∂X and let Ω be the canonical Riemannian volume form on X itself. Let $\mathbf{n}$ be the unit outward normal vector field to the boundary, and let ξ be a C^1 vector field on X, with compact support. Then*

$$\int_X (\mathrm{div}_{\Omega}\, \xi)\Omega = \int_{\partial X} \langle \mathbf{n}, \xi \rangle \omega.$$

The next thing is to show that the map d^* from §2 is the adjoint for a scalar product defined by integration. First we expand slightly the formalism of d^* for this application. Recall that for any vector field ξ, the divergence of ξ is defined by the property

(1) $$d(\mathrm{vol}_g \circ \xi) = (\mathrm{div}\,\xi)\mathrm{vol}_g.$$

Note the trivial derivation formula for a function φ:

(2) $$\mathrm{div}(\varphi \xi) = \varphi\, \mathrm{div}\, \xi + (d\varphi)(\xi).$$

If λ is a 1-form, i.e. in $\Gamma L^1(TX) = \mathcal{A}^1(X)$, we have the corresponding vector field $\xi_\lambda = \lambda^\vee$ uniquely determined by the condition that

$$\langle \xi_\lambda, \eta \rangle_g = \lambda(\eta) \qquad \text{for all vector fields } \eta.$$

For a 1-form λ, we define the operator

$$d^*: \mathcal{A}^1(X) \to \mathcal{A}^0(X) = \text{Fu}(X) \qquad \text{by} \qquad d^*\lambda = -\text{div } \xi_\lambda,$$

so by (1),

$$(3) \qquad\qquad (d^*\lambda) \text{ vol}_g = d(\text{vol}_g \circ \xi_\lambda).$$

We get a formula analogous to (2) for d^*, namely

$$(4) \qquad\qquad d^*(\varphi\lambda) = \varphi d^*\lambda - \langle d\varphi, \lambda \rangle.$$

Indeed, $d^*(\varphi\lambda) = -\text{div } \xi_{\varphi\lambda} = -\text{div}(\varphi\xi_\lambda) = -\varphi \text{ div } \xi_\lambda - (d\varphi)(\xi_\lambda)$ by (2), which proves the formula.

Let $\lambda, \omega \in \mathcal{A}^1(TX)$. We define the **scalar product** via duality

$$\langle \lambda, \omega \rangle_g = \langle \xi_\lambda, \xi_\omega \rangle_g.$$

Then for a function φ we have the formula

$$(5) \qquad\qquad \langle d\varphi, \lambda \rangle_g \text{ vol}_g = (\varphi d^*\lambda) \text{ vol}_g - d(\text{vol}_g \circ \varphi\xi_\lambda).$$

Indeed,

$$\langle d\varphi, \lambda \rangle_g \text{ vol}_g = [\varphi \, d^*\lambda - d^*(\varphi\lambda)] \text{ vol}_g \qquad \text{by (4)}$$
$$= (\varphi d^*\lambda) \text{ vol}_g - d(\text{vol}_g \circ \varphi\xi_\lambda) \quad \text{by (3)}$$

thus proving (5). Note that the congruence of the two forms $\langle d\varphi, \lambda \rangle_g \text{ vol}_g$ and $(\varphi d^*\lambda) \text{ vol}_g$ modulo exact forms is significant, and is designed for Proposition 3.3 below.

Observe that the scalar product between two forms above is a function, which when multiplied by the volume form vol_g may be integrated over X. Thus we define the **global scalar product** on 1-forms with compact support to be

$$\langle \lambda, \omega \rangle_{(X, g)} = \langle \lambda, \omega \rangle_X = \int_X \langle \lambda, \omega \rangle_g \text{ vol}_g.$$

Applying Stokes' theorem, we then find:

Proposition 3.3. *Let (X, g) be a Riemannian manifold, oriented and without boundary. Then d^* is the adjoint of d with respect to the global scalar product, i.e.*

$$\langle d\varphi, \lambda \rangle_X = \langle \varphi, d^*\lambda \rangle_X.$$

We define the **Laplacian** (operating on functions) to be the operator

$$\Delta = d^*d.$$

For a manifold with boundary, we define the **normal derivative** of a function φ to be the function *on the boundary* given by

$$\partial_{\mathbf{n}}\, \varphi = \langle \mathbf{n}, \xi_{d\varphi} \rangle_g = \langle \mathbf{n}, \operatorname{grad}_g \varphi \rangle_g.$$

Theorem 3.4 (Green's Formula). *Let (X, g) be an oriented Riemannian manifold possibly with boundary, and let φ, ψ be functions on X with compact support. Let ω be the canonical volume form associated with the induced metric on the boundary. Then*

$$\int_X (\varphi\Delta\psi - \psi\Delta\varphi)\, \mathrm{vol}_g = -\int_{\partial X} (\varphi\partial_{\mathbf{n}}\psi - \psi\partial_{\mathbf{n}}\varphi)\omega.$$

Proof. From formula (4) we get

$$d^*(\varphi\, d\psi) = \varphi\Delta\psi - \langle d\varphi, d\psi \rangle_g,$$

whence

$$\varphi\Delta\psi - \psi\Delta\varphi = d^*(\varphi\, d\psi) - d^*(\psi\, d\varphi)$$
$$= -\operatorname{div}(\varphi\, d\psi) + \operatorname{div}(\psi\, d\varphi).$$

We apply Theorem 3.2 to conclude the proof.

Remark. Of course, if X has no boundary in Theorem 3.7, then the integral on the left side is equal to 0.

Corollary 3.5 (E. Hopf). *Let X be an oriented Riemannian manifold without boundary, and let f be a C^2 function on X with compact support, such that $\Delta f \geqq 0$. Then f is constant. In particular, every harmonic function with compact support is constant.*

Proof. By Green's formula we get

$$\int_X \Delta f\, \mathrm{vol}_g = 0.$$

Since $\Delta f \geqq 0$, it follows that in fact $\Delta f = 0$, so we are reduced to the harmonic case. We now apply Green's formula to f^2, and get

$$0 = \int_X \Delta f^2 \, \text{vol}_g = \int_X 2f\Delta f \, \text{vol}_g - \int_X 2(\text{grad } f)^2 \, \text{vol}_g.$$

Hence $(\text{grad } f)^2 = 0$ because $\Delta f = 0$, and finally grad $f = 0$, so $df = 0$ and f is constant, thus proving the corollary.

X, §4. CAUCHY'S THEOREM

It is possible to define a complex analytic (analytic, for short) manifold, using open sets in $\mathbf{C}^n$ and charts such that the transition mappings are analytic. Since analytic amps are C^∞, we see that we get a C^∞ manifold, but with an additional structure, and we call such a manifold **complex analytic**. It is verified at once that the analytic charts of such a manifold define an orientation. Indeed, under a complex analytic change of charts, the Jacobian changes by a complex number times its complex conjugate, so changes by a positive real number.

If $z_1, \ldots, z_n$ are the complex coordinates of $\mathbf{C}^n$, then

$$(z_1, \ldots, z_n, \bar{z}_1, \ldots, \bar{z}_n)$$

can be used as C^∞ local coordinates, viewing $\mathbf{C}^n$ as $\mathbf{R}^{2n}$. If $z_k = x_k + iy_k$, then

$$dz_k = dx_k + i \, dy_k \qquad \text{and} \qquad d\bar{z}_k = dx_k - i \, dy_k.$$

Differential forms can then be expressed in terms of wedge products of the dz_k and $d\bar{z}_k$. For instance

$$dz_k \wedge d\bar{z}_k = 2i \, dy_k \wedge dx_k.$$

The complex standard expression for a differential form is then

$$\omega(z) = \sum_{(i, j)} \varphi_{(i, j)}(z) \, dz_{i_1} \wedge \cdots \wedge dz_{i_r} \wedge d\bar{z}_{j_1} \wedge \cdots \wedge d\bar{z}_{j_s}.$$

Under an analytic change of coordinates, one sees that the numbers r and s remain unchanged, and that if $s = 0$ in one analytic chart, then $s = 0$ in any other analytic chart. Similarly for r. Thus we can speak of a form of type (r, s). A form is said to be **analytic** if $s = 0$, that is if it is of type $(r, 0)$.

We can decompose the exterior derivative d into two components. Namely, we note that if ω is of type (r, s), then $d\omega$ is a sum of forms of type $(r+1, s)$ and $(r, s+1)$, say

$$d\omega = (d\omega)_{(r+1, s)} + (d\omega)_{(r, s+1)}.$$

We define

$$\partial\omega = (d\omega)_{(r+1, s)} \quad \text{and} \quad \bar{\partial}\omega = (d\omega)_{(r, s+1)}.$$

In terms of local coordinates, it is then easy to verify that if ω is decomposable, and is expressed as

$$\omega(z) = \varphi(z)dz_{i_1} \wedge \cdots \wedge dz_{i_r} \wedge d\bar{z}_{j_1} \wedge \cdots \wedge d\bar{z}_{j_s} = \varphi\tilde{\omega},$$

then

$$\partial\omega = \sum \frac{\partial\varphi}{\partial z_k} dz_k \wedge \tilde{\omega}.$$

and

$$\bar{\partial}\omega = \sum \frac{\partial\varphi}{\partial \bar{z}_k} d\bar{z}_k \wedge \tilde{\omega}.$$

In particular, we have

$$\frac{\partial}{\partial z_k} = \frac{1}{2}\left(\frac{\partial}{\partial x_k} - i\frac{\partial}{\partial y_k}\right) \quad \text{and} \quad \frac{\partial}{\partial \bar{z}_k} = \frac{1}{2}\left(\frac{\partial}{\partial x_k} + i\frac{\partial}{\partial y_k}\right).$$

(*Warning*: Note the position of the plus and minus signs in these expressions.)

Thus we have

$$d = \partial + \bar{\partial},$$

and operating with ∂ or $\bar{\partial}$ follows rules similar to the rules for operating with d.

Note that f is analytic if and only if $\bar{\partial}f = 0$. Similarly, we say that a differential form is **analytic** if in its standard expression, the functions $\varphi_{(i, j)}$ are analytic and the form is of type $(r, 0)$, that is there are no $d\bar{z}_j$ present. Equivalently, this amounts to saying that $\bar{\partial}\omega = 0$. The following extension of Cauchy's theorem to several variables is due to Martinelli.

We let $|z|$ be the euclidean norm,

$$|z| = (z_1\bar{z}_1 + \cdots + z_n\bar{z}_n)^{1/2}.$$

Theorem 4.1 (Cauchy's Theorem). *Let f be analytic on an open set in $\mathbb{C}^n$ containing the closed ball of radius R centered at a point ζ. Let*

$$\omega_k(z) = dz_1 \wedge \cdots \wedge dz_n \wedge d\bar{z}_1 \wedge \cdots \wedge \widehat{d\bar{z}_k} \wedge \cdots \wedge d\bar{z}_n$$

and

$$\omega(z) = \sum_{k=1}^{n} (-1)^k \bar{z}_k \omega_k(z).$$

Let S_R be the sphere of radius R centered at ζ. Then

$$f(\zeta) = \epsilon(n) \frac{(n-1)!}{(2\pi i)^n} \int_{S_R} \frac{f(z)}{|z - \zeta|^{2n}} \omega(z - \zeta)$$

where $\epsilon(n) = (-1)^{n(n+1)/2}$.

Proof. We may assume $\zeta = 0$. First note that

$$\bar{\partial}\omega(z) = \sum_{k=1}^{n} (-1)^k d\bar{z}_k \wedge \omega_k(z) = (-1)^{n+1} n \, dz \wedge d\bar{z},$$

where $dz = dz_1 \wedge \cdots \wedge dz_n$ and similarly for $d\bar{z}$. Next, observe that if

$$\psi(z) = \frac{f(z)}{|z|^{2n}} \omega(z),$$

then

$$d\psi = 0.$$

This is easily seen. On the one hand, $\partial\psi = 0$ because ω already has $dz_1 \wedge \cdots \wedge dz_n$, and any further dz_i wedged with this gives 0. On the other hand, since f is analytic, we find that

$$\bar{\partial}\psi(z) = f(z) \bar{\partial}\left(\frac{\omega(z)}{|z|^{2n}} \right) = 0$$

by the rule for differentiating a product and a trivial computation.

Therefore, by Stokes' theorem, applied to the annulus between two spheres, for any r with $0 < r \le R$ we get

$$\int_{S_R} \psi - \int_{S_r} \psi = 0,$$

or in other words,

$$\int_{S_R} f(z) \frac{\omega(z)}{|z|^{2n}} = \int_{S_r} f(z) \frac{\omega(z)}{|z|^{2n}}$$

$$= \frac{1}{r^{2n}} \int_{S_r} f(z) \omega(z).$$

Using Stokes' theorem once more, and the fact that $\partial\omega = 0$, we see that this is

$$= \frac{1}{r^{2n}} \int_{B_r} \bar{\partial}(f\omega) = \frac{1}{r^{2n}} \int_{B_r} f \bar{\partial}\omega.$$

We can write $f(z) = f(0) + g(z)$, where $g(z)$ tends to 0 as z tends to 0. Thus in taking the limit as $r \to 0$, we may replace f by $f(0)$. Hence our last expression has the same limit as

$$f(0)\frac{1}{r^{2n}} \int_{B_r} \bar{\partial}\omega = f(0)\frac{1}{r^{2n}} \int_{B_r} (-1)^{n+1} n \, dz \wedge d\bar{z}.$$

But

$$dz \wedge d\bar{z} = (-1)^{n(n-1)/2} \, i^n \, 2^n \, dy_1 \wedge dx_1 \wedge \cdots \wedge dy_n \wedge dx_n.$$

Interchanging dy_k and dx_k to get the proper orientation gives another contribution of $(-1)^n$, together with the form giving Lebesgue measure. Hence our expression is equal to

$$f(0)(-1)^{n(n+1)/2} n(2i)^n \frac{1}{r^{2n}} V(B_r),$$

where $V(B_r)$ is the Lebesgue volume of the ball of radius r in $\mathbf{R}^{2n}$, and is classically known to be equal to $\pi^n r^{2n}/n!$. Thus finally we see that our expression is equal to

$$f(0)(-1)^{n(n+1)/2} \frac{(2\pi i)^n}{(n-1)!}.$$

This proves Cauchy's theorem.

X, §5. THE RESIDUE THEOREM

Let f be an analytic function in an open set U of $\mathbf{C}^n$. The set of zeros of f is called a **divisor**, which we denote by $V = V_f$. In the neighborhood of a regular point a, that is a point where $f(a) = 0$ but some complex partial derivative of f is not zero, the set V is a complex submanifold of U. In fact, if, say, $D_n f(a) \neq 0$, then the map

$$(z_1, \ldots, z_n) \mapsto (z_1, \ldots, z_{n-1}, f(z))$$

gives a local analytic chart (analytic isomorphism) in a neighborhood of a. Thus we may use f as the last coordinate, and locally V is simply obtained by the projection on the set $f = 0$. This is a special case of the complex analytic inverse function theorem.

It is always true that the function $\log |f|$ is locally in $\mathscr{L}^1$. We give the proof only in the neighborhood of a regular point a. In this case, we can change f by a chart (which is known as a change-of-variable formula), and we may therefore assume that $f(z) = z_n$. Then $\log|f| = \log |z_n|$, and the Lebesgue integral decomposes into a simple product integral, which reduces our problem to the case of one variable, that is to the fact that $\log|z|$ is locally integrable near 0 in the ordinary complex plane. Writing $z = re^{i\theta}$, our assertion is obvious since the function $r \log r$ is locally integrable near 0 on the real line.

Note. In a neighborhood of a singular point the fastest way and formally clearest, is to invoke Hironaka's resolution of singularities, which reduces the question to the non-singular case.

For the next theorem, it is convenient to let

$$d^c = \frac{1}{4\pi i}(\partial - \bar{\partial}).$$

Note that

$$dd^c = \frac{i}{2\pi} \partial \bar{\partial}.$$

The advantage of dealing with d and d^c is that they are real operators.

The next theorem, whose proof consists of repeated applications of Stokes' theorem, is due to Poincaré. It relates integration in V and U by a suitable kernel.

Theorem 5.1 (Residue Theorem). *Let f be analytic on an open set U of $\mathbb{C}^n$ and let V be its divisor of zeros in U. Let ψ be a C^∞ form with compact support in U, of degree $2n - 2$ and type $(n - 1, n - 1)$. Then*

$$\int_V \psi = \int_U \log |f|^2 \, dd^c \psi.$$

(As usual, the integral on the left is the integral of the restriction of ψ to V, and by definition, it is taken over the regular points of V.)

Proof. Since ψ and $dd^c\psi$ have compact support, the theorem is local (using partitions of unity). We give the proof only in the neighborhood of a regular point. Therefore we may assume that U is selected sufficiently small so that every point of the divisor of f in U is regular, and such that, for small ϵ, the set of points

$$U_\epsilon = \{z \in U, \, |f(z)| \geq \epsilon\}$$

is a submanifold with boundary in U. The boundary of U_ϵ is then the set of points z such that $|f(z)| = \epsilon$. (Actually to make this set a submanifold we only need to select ϵ to be a regular value, which can be done for arbitrarily small ϵ by Sard's theorem.) For convenience we let S_ϵ be the boundary of U_ϵ, that is the set of points z such that $|f(z)| = \epsilon$.

Since $\log |f|$ is locally in $\mathscr{L}^1$, it follows that

$$\int_{U_\epsilon} \log |f| \, dd^c\psi = \lim_{\epsilon \to 0} \int_{U_\epsilon} \log |f| \, dd^c\psi.$$

Using the trivial identity

$$d(\log |f| \, d^c\psi) = d \log |f| \wedge d^c\psi + \log |f| \, dd^c\psi,$$

we conclude by Stokes' theorem that this limit is equal to

$$\lim_{\epsilon \to 0} \left[\int_{S_\epsilon} \log |f| \, d^c\psi - \int_{U_\epsilon} d \log |f| \wedge d^c\psi \right].$$

The first integral under the limit sign approaches 0. Indeed, we may assume hat $f(z) = z_n = re^{i\theta}$. On S_ϵ we have $|f(z)| = \epsilon$, so $\log |f| = \log \epsilon$. There exist forms ψ_1, ψ_2 in the first $n - 1$ variables such that

$$d^c\psi = \psi_1 \wedge dz_n + \psi_2 \wedge d\bar{z}_n,$$

and the restriction of dz_n to S_ϵ is equal to

$$\epsilon i e^{i\theta} \, d\theta,$$

with a similar expression for $d\bar{z}_n$. Hence our boundary integral is of type

$$\epsilon \log \epsilon \int_{S_\epsilon} \omega,$$

where ω is a bounded form. From this it is clear that the limit is 0.

Now we compute the second integral. Since ψ is assumed to be of type $(n - 1, n - 1)$ it follows that for any function g,

$$\partial g \wedge \partial \psi = 0 \qquad \text{and} \qquad \bar{\partial} g \wedge \bar{\partial} \psi = 0.$$

Replacing d and d^c by their values in terms of ∂ and $\bar{\partial}$, it follows that

$$-\int_{U_\epsilon} d \log |f| \wedge d^c\psi = \int_{U_\epsilon} d^c \log |f| \wedge d\psi.$$

We have

$$d(d^c \log |f| \wedge \psi) = dd^c \log |f| \wedge \psi - d^c \log |f| \wedge d\psi.$$

Furthermore dd^c is a constant times $\partial \bar{\partial}$, and $dd^c \log |f|^2 = 0$ in any open set where $f \neq 0$, because

$$\partial \bar{\partial} \log |f|^2 = \partial \bar{\partial} (\log f + \log \bar{f}) = 0$$

since $\partial \log \bar{f} = 0$ and $\bar{\partial} \log f = 0$ by the local analyticity of $\log f$. Hence we obtain the following values for the second integral by Stokes:

$$\int_{U_\epsilon} d^c \log |f|^2 \wedge d\psi = \int_{S_\epsilon} d^c \log |f|^2 \wedge \psi.$$

Since

$$d^c \log |f|^2 = -\frac{i}{4\pi} (\partial - \bar{\partial})(\log f + \log \bar{f})$$

$$= -\frac{i}{4\pi} \left(\frac{dz_n}{z_n} - \frac{d\bar{z}_n}{\bar{z}_n} \right)$$

(always assuming $f(z) = z_n$), we conclude that if $z_n = re^{i\theta}$, then the restriction of $d^c \log |f|^2$ to S_ϵ is given by

$$\mathrm{res}_{S_\epsilon} d^c \log f = \frac{d\theta}{2\pi}.$$

Now write ψ in the form

$$\psi = \psi_1 + \psi_2$$

where ψ_1 contains only dz_j, $d\bar{z}_j$ for $j = 1, \ldots, n-1$ and ψ_2 contains dz_n or $d\bar{z}_n$. Then the restriction of ψ_2 to S_ϵ contains $d\theta$, and consequently

$$\int_{S_\epsilon} d^c \log |f|^2 \wedge \psi = \int_{S_\epsilon} \frac{d\theta}{2\pi} \wedge (\psi_1 | S_\epsilon).$$

The integral over S_ϵ decomposes into a product integral, we respect to the first $n-1$ variables, and with respect to $d\theta$. Let

$$\int^{(n-1)} \psi_1(z) | S_\epsilon = g(z_n).$$

Then simply by the continuity of g we get

$$\lim_{\epsilon \to 0} \frac{1}{2\pi} \int_0^{2\pi} g(\epsilon e^{i\theta}) \, d\theta = g(0).$$

Hence

$$\lim_{\epsilon \to 0} \int_{S_\epsilon} \frac{d\theta}{2\pi} \wedge (\psi_1 \,|\, S_\epsilon) = \int_{z_n=0} \psi_1.$$

But the restriction of ψ_1 to the set $z_n = 0$ (which is precisely V) is the same as the restriction of ψ to V. This proves the residue theorem.

Correction to: Introduction to Differentiable Manifolds

Serge Lang

The title of this book was incorrectly captured as Introduction to Differential Manifolds. The correct title should be Introduction to Differentiable Manifolds. This has been corrected.

The updated online version of this book can be found at
https://doi.org/10.1007/b97450_11

C1

Bibliography

[Ab 62]   R. ABRAHAM, *Lectures of Smale on Differential Topology*, Columbia University, 1962

[AbM 78]   R. ABRAHAM and J. MARSDEN, *Foundations of Mechanics*, second edition, Benjamin-Cummings, 1978

[APS 60]   W. AMBROSE, R.S. PALAIS, and I.M. SINGER, Sprays, *Acad. Brasileira de Ciencias* **32** (1960) pp. 163–178

[BGM 71]   M. BERGER, P. GAUDUCHON, and E. MAZET, *Le Spectre d'une Variété Riemannienne*, Lecture Notes in Mathematics **195**, Springer-Verlag, 1971

[Be 78]   A.L. BESSE, *Manifolds all of whose geodesics are closed*, Ergebnisse der Math. **93**, Springer-Verlag, 1978

[Bo 60]   R. BOTT, *Morse Theory and its Applications to Homotopy Theory*, Lecture Notes by van de Ven, Bonn, 1960

[BoT 82]   R. BOTT and L. TU, *Differential Forms in Algebraic Topology*, Graduate Texts in Mathematics **82**, Springer-Verlag, 1982

[Bou 68]   N. BOURBAKI, *General Topology*, Addison-Wesley, 1968 (translated from the French, 1949)

[Bou 69]   N. BOURBAKI, *Fasicule de Résultats des Variétés*, Hermann, 1969

[doC 92]   M.P. do CARMO, *Riemannian Geometry*, Birkhaüser, 1992

[Ca 28]   E. CARTAN, *Leçons sur la Géometrie des Espaces de Riemann*, Gauthier-Villars, Paris, 1928, second edition 1946

[Do 68]   P. DOMBROWSKI, Krümmungsgrössen gleichungsdefinierte Untermannigfaltigkeiten, *Math. Nachr.* **38** (1968) pp. 133–180

[GHL 87/93] S. GALLOT, D. HULIN, and J. LAFONTAINE, *Riemannian Geometry*, second edition, Springer-Verlag, 1987, 1993 (corrected printing)

[God 58]   R. GODEMENT, *Topologie algébrique et théorie des faisceaux*, Hermann, 1958

[Gr 61]   W. GRAEUB, Liesche Grupen und affin zusammenhangende Man-
               nigfaltigkeiten, *Acta Math.* (1961) pp. 65–111

[GrKM 75]   D. GROMOLL, W. KLINGENBERG, and W. MEYER, *Riemannsche
               Geometrie im Grössen*, Lecture Notes in Mathematics **55**,
               Springer-Verlag, 1975

[Gu 91]   P. GUNTHER, Huygens' Principle and Hadamard's Conjecture,
               Math. Intelligencer **13** No. 2 (1991) pp. 56–63

[He 61]   S. HELGASON, Some remarks for the exponential mapping for an
               affine connection, *Math. Scand.* **9** (1961) pp. 129–146

[He 78]   S. HELGASON, *Differential geometry, Lie groups, and symmetric
               spaces*, Academic Press, 1978

[He 84]   S. HELGASON, Wave Equations on Homogeneous Spaces, Lecture
               Notes in Mathematics 1077, Springer-Verlag 1984, pp. 254–287

[Ir 70]   M.C. IRWIN, On the stable manifold theorem, *Bull. London Math.
               Soc.* (1970) pp. 68–70

[Ke 55]   J. KELLEY, *General Topology*, Van Nostrand, 1955

[Kl 83/95]   W. KLINGENBERG, *Riemannian Geometry*, de Gruyter, 1983; second
               edition 1995

[Ko 57]   S. KOBAYASHI, Theory of Connections, *Annali di Mat.* **43** (1957)
               pp. 119–194

[Ko 87]   S. KOBAYASHI, *Differential Geometry of Complex Vector Bundles*
               Iwanami Shoten and Princeton University Press, 1987

[KoN 63]   S. KOBAYASHI and K. NOMIZU, *Foundations of Differential Geometry*
               I, Wiley, 1963 and 1969

[KoN 69]   S. KOBAYASHI and K. NOMIZU, *Foundations of Differential Geometry*
               II, Wiley, 1969

[La 62]   S. LANG, *Introduction to Differentiable Manifolds*, Addison-Wesley,
               1962

[La 71]   S. LANG, *Differential Manifolds*, Addison-Wesley, 1971; Springer-
               Verlag, 1985

[La 83/97]   S. LANG, *Undergraduate Analysis*, Springer-Verlag, 1983; Second
               Edition, 1997

[La 93]   S. LANG, *Real and Functional Analysis*, third edition, Springer-
               Verlag, 1993

[La 99]   S. LANG, *Fundamentals of Differential Geometry*, Springer-Verlag,
               1999

[La 02]   S. LANG, *Algebra* (Third revised edition), Springer-Verlag, 2002

[LoS 68]   L. LOOMIS and S. STERNBERG, *Advanced Calculus*, Addison-Wesley,
               1968

[Lo 69]   O. LOOS, *Symmetric spaces* I and II, Benjamin, 1969

[Mar 74]   J. MARSDEN, *Applications of Global Analysis to Mathematical
               Physics*, Publish or Perish, 1974; reprinted in "Berkeley Mathe-
               matics Department Lecture Note Series", available from the
               UCB Math. Dept.

[MaW 74]   J. MARSDEN and A. WEINSTEIN, Reduction of symplectic manifolds
               with symmetry, *Rep. Math. Phys.* **5** (1974) pp. 121–130

[Maz 61]   B. MAZUR, Stable equivalence of differentiable manifolds, *Bull. Amer. Math. Soc.* **67** (1961) pp. 377–384

[Mi 58]   J. MILNOR, *Differential Topology*, Princeton University Press, 1968

[Mi 59]   J. MILNOR, *Der Ring der Vectorraumbundel eines topologischen Raümes*, Bonn, 1959

[Mi 61]   J. MILNOR, *Differentiable Structures*, Princeton University Press, 1961

[Mi 63]   J. MILNOR, *Morse Theory*, Ann. Math. Studies **53**, Princeton University Press, 1963

[Mo 61]   J. MOSER, A new technique for the construction of solutions for nonlinear differential equations, *Proc. Nat. Acad. Sci. USA* **47** (1961) pp. 1824–1831

[Mo 65]   J. MOSER, On the volume element of a manifold, *Trans. Amer. Math. Soc.* **120** (1965) pp. 286–294

[Nas 56]   J. NASH, The embedding problem for Riemannian manifolds, *Ann. of Math.* **63** (1956) pp. 20–63

[O'N 97]   B. O'NEILL, *Elementary Differential Geometry*, second edition, Academic Press, 1997

[Pa 68]   R. PALAIS, *Foundations of Global Analysis*, Benjamin, 1968

[Pa 69]   R. PALAIS, The Morse lemma on Banach spaces, *Bull. Amer. Math. Soc.* **75** (1969) pp. 968–971

[PaS 64]   R. PALAIS and S. SMALE, A generalized Morse theory, *Bull. Amer. Math. Soc.* **70** (1964) pp. 165–172

[Pro 70]   *Proceedings of the Conference on Global Analysis, Berkeley, Calif. 1968*, AMS, 1970

[Sm 63]   S. SMALE, Stable manifolds for differential equations and diffeomorphism, *Ann. Scuola Normale Sup. Pisa* **III** Vol. XVII (1963) pp. 97–116

[Sm 64]   S. SMALE, Morse theory and a non-linear generalization of the Dirichlet problem, *Ann. of Math.* **80** No. 2 (1964) pp. 382–396

[Sm 67]   S. SMALE, Differentiable dynamical systems, *Bull. Amer. Math. Soc.* **73**, No. 6 (1967) pp. 747–817

[Sp 70/79] M. SPIVAK, *Differential Geometry* (5 volumes), Publish or Perish, 1970–1979

[Ste 51]   N. STEENROD, *The Topology of Fiber Fundles*, Princeton University Press, 1951

[Str 64/83] S. STERNBERG, *Lectures on Differential Geometry*, Prentice-Hall, 1964; second edition, Chelsea, 1983

[Wel 80]   R. WELLS, *Differential Analysis on Complex Manifolds*, Graduate Texts in Mathematics **65**, Springer-Verlag, 1980

[Wu 65]   H.-S. WU, Two remarks on sprays, *J. Math. Mech.* **14** No. 5 (1965) pp. 873–880

Index

Printed in the United States
by Baker & Taylor Publisher Services